GARDENING ON CLAY AND LIME

ALSO BY THE SAME AUTHOR

A Flower for Every Day
An all the year Garden
Carefree Gardening
Cottage Garden Flowers
Gardening in the Shade
Ground Cover Plants
We Made a Garden

GARDENING ON CLAY AND LIME

by

MARGERY FISH

DAVID & CHARLES : NEWTON ABBOT

7153 4862 0

Set in 11/13 pt Plantin
and printed in Great Britain
by Bristol Typesetting Company Limited Bristol
for David & Charles (Publishers) Limited
South Devon House Newton Abbot Devon

CONTENTS

Acknowledgments

Grateful thanks are due to Chris Brickell for his help in checking the names of plants; to Miss Hodgman and Derek Tilley for supplying the photographs, and to Mrs Jean Goscomb for typing the manuscript.

List of Illustrations

PLATES

(Unless otherwise stated these are the copyright of Miss M. E. Hodgman.)

INTRODUCTORY NOTE

Margery Fish died on 24 March 1969 shortly after she had completed the manuscript of this book.

Her death came as a great shock not only to her many friends but to countless gardeners throughout the world who, though they had never met her, heard her lecture, or even seen her garden, felt that they had come to know her intimately through her books and articles. For she had a unique ability to inject her enthusiasm and the warmth of her tremendous personality into her writing.

Anyone who was lucky enough to have visited her at East Lambrook Manor knows what an exciting experience for any gardener this invariably was, and how hearty the welcome which she gave even to those whom she had never previously met but whose interest in plants had driven them to seek her out at the heart of a maze of narrow country lanes.

Her influence on horticulture during her lifetime was immense, both in her whole approach to gardening, emphasising as she did the importance of ground cover and perennials in order to provide all-the-year interest while at the same time keeping the effort involved to a minimum, and also in relation to the plants which she restored to popularity, many of them previously in danger of virtual extinction. As a close friend said in a tribute paid shortly after her death: 'She was a luminary among us, and she threw a light for us on a whole world of plants which my generation didn't know—the plants of the old-fashioned cottage garden.' Perhaps these two aspects of her work are more closely related than even she realised.

Her garden at East Lambrook Manor remains as a living memorial to her for all to see, and there her ideas were put into effect in the idyllic setting which, during the last thirty years of her life, she had worked so hard to create from a derelict farmyard. One of the most usual remarks made by visitors to the garden is: 'What wonderful soil it must be—everything grows so well here!' And that is the answer to the defeatism so often conjured up by the very thought of gardening on clay and lime.

H.B-C.

I

DEALING WITH CLAY

Wonderful results can be obtained from clay soil and a garden on heavy soil can be one of the most beautiful. Certain principles have to be observed and one always must work with the soil and not against it.

One of the first things I learnt when I started gardening on my heavy clay soil was never to walk on it when it was wet. The tiny particles of soil adhere so closely together that very little air circulates in it, and putting weight on the soil when it is wet increases this condition. This was first brought home to me when I started planting things out in rows. Every time I walked between the rows I had to fork up the soil to obliterate my footsteps; otherwise it turned green and sour, and became as hard as concrete when it dried. Obviously I couldn't stop all gardening when the soil was wet, particularly in a county like Somerset which has a great deal of rain at times. Standing on a plank is a help but it is even better if all work can be done from a stone path. As far as possible all my flower beds are fairly narrow so that I can reach the middle from one side or the other. The border against the wall that shields me from the road had to be made wider if a good effect was to be had, so I laid a narrow stone path towards the back of the bed, from which I can tend the narrow space between this path and the wall, and a good 3—4 ft at the back of the wide bed. Instead of planting out nursery plants in long rows, I had beds made of a convenient size—3 ft across with breeze block paths all round them. Although they do not hold as many plants as the long rows did, they produce better results. There is no need to walk on the beds at all so the soil keeps loose and open, and the plants nearest the paths are particularly fine because they can get their roots under stones, which most plants prefer.

In other parts of the garden where I plant out rooted cuttings and seedlings, I have adopted the block system, but with irregular enclosures instead of rectangular ones, to give a more attractive, informal look. The main reason is, of course, to separate the different varieties, but the layout has other advantages, one being that the size of the enclosures can be varied according to the number of plants in each. I am convinced that the majority of plants are happier if growing close to others, so on this score the block system is good, always assuming that each plant is given sufficient room to develop.

Every aspect of these irregular plantings is informal, including the stones used to make the paths. As the enclosures are odd shapes it would not be possible to use uniform rectangular stones, such as the concrete blocks which form the paths between the nursery beds. Pieces of paving or of concrete slabs are suitable. I am lucky because I can get plenty of broken hamstone tiles, which, though thin, are heavy enough to sink into the ground that has been dug to a loose tilth. The stone tiles that roof many houses in Somerset are quite large, sometimes about 30 in wide and 20 in deep, and are hung on nails on the roof. They are seldom used now but most builders have heaps of broken ones, usually about 1 ft across, sometimes longer, and all very suitable for irregular narrow paths. I prefer them to thicker stones, which have to be sunk and carefully levelled, and it is not difficult to lift the stones to alter the shapes of the enclosures or to weed between them.

Larger thicker stones are needed if permanent paths are necessary in these irregular-shaped plantings. I have no trouble in one large shady patch, which does not get too dry, but in a south-facing part of the garden, which has a slight slope, one or two wider permanent paths have had to be made. A large number of people walk through the area, studying the silver plants, pinks and euphorbias that are grown here, and the stone tiles are kicked about on the top of the soil which is as hard as concrete in summer. Stepping stones are useful in wide beds because with plants of all heights a small stone path through the middle would be too obvious. A flat stone here and there on which to stand enables the gardener to carry on with his work whatever the state of the soil.

Any lawn gets bare if there is continual traffic along a certain route and on clay soil recovery takes longer because the ground becomes so much harder. On any lawn where people take the shortest route it is essential to sink large, flat, solid stones far enough below the level of the grass not to affect the mower. For those who do not live in a stone county, concrete slabs very soon weather to a neutral shade, and to hasten the process they can be washed with water in which a small bag of manure has been soaked.

HANDLING THE SOIL

The orthodox methods of digging and double digging may work in some kinds of clay if done with extreme care by a skilled worker, but in the case of a novice I would recommend all cultivation to be confined to the top 12 in or thereabouts, as the deeper one digs the stickier and more impenetrable the soil becomes.

When growing vegetables different techniques are needed. Double digging is necessary for root vegetables, particularly those with long tapering roots such as parsnips and carrots. It is usual to open a trench for runner beans; the lower strata of soil should be well broken up with a fork and then filled with compost. When I grew vegetables I dug my trenches early and filled them with all the material that would normally have gone on the compost heap, until it was time to plant. Then I covered the compost with the top soil that had been carefully heaped nearby.

Before we made our garden the part that is now terraced was a gentle slope, so to make level beds and paths a great deal of soil had to be taken away. I tried very hard to keep the precious top soil safe; it was carefully removed to a depth of at least 15 in and left in piles, but in spite of infinite care some subsoil was in the top layer when the beds were finished. It took me many years to get rid of it, and even now if I dig too deeply small lumps appear on the surface.

Young people who buy new houses are often faced with the problem of subsoil being left by the builder on top of the good soil, and this is enough to put the owners off gardening for life. After a few years' experience it is easy to distinguish the difference at a glance, but to the beginner soil is just soil and he wonders why his plants take so long to settle down.

The small lumps of subsoil do not break down and become incorporated in the top soil, but produce air pockets below the surface so that the roots of the plants often have nothing on which to take hold.

Very early in my gardening days I adopted the technique which I still follow, of planting everything in holes filled with

is good too. When black with age, it has an excellent effect on the soil but it is always recommended to avoid new sawdust as there is a reduction in the available nitrogen for plant use due to bacterial action on the fresh sawdust. There is also said to be the danger that it encourages eel-worm, though I do not know that this has been proved. When I had more time I collected leaves and leaf-mould wherever I could find it. A friend in Devonshire owned some woodland and I could have as much leaf-mould as I could carry. It was under beeches, deep and moist, and easy to dig up, but sometimes the deposit is shallow and tree roots are near the surface, so that one has to scratch away for a long time without getting a great deal. This ready-rotted leaf-mould I used at once as I always had plants that seemed to be crying out for it, particularly primroses. Some people sieve it but I would never do that except perhaps for seed sowing. Now I have turned to peat as easier and quicker to acquire, but it lacks the coarse texture and pleasant woodland smell of leaf-mould. At one time I often found vine weevils—fat white grubs that eat roots—when I dug up primulas and primroses and it was suggested that they may have been introduced in the leaf-mould, but again one has no proof.

If leaves are collected each year it is always recommended that they should be stacked by themselves and not put into a compost heap. For one thing some leaf-types take longer than other vegetation to rot down. For years I was able to have as many bags of leaves as I cared to collect. Friends with a big garden had large beech trees along the drive to their house and when the leaves fell gardeners swept them into piles to be collected either for their own use or by friends. I kept some big sacks for this yearly bonus, and always in autumn had several sacks in the car in case I passed through country where there were a great many trees, beech and oak in particular. Sometimes at night the heaps of leaves on lonely roads were more than I could resist and I remember once stopping after leaving a cocktail party to fill my sacks by the light of the car lamps. But I quite forgot that other guests would be leaving by the same route and after several stopped to ask if I was all right I decided this was not a thing to do quite so publicly.

Anyone who has collected fallen leaves knows they take a long time to rot down. I never use activators of any kind and my leaves were not really rotted after two years. In a large garden it is usually possible to put them in a pile, but in a small garden something tidier is needed for the wind blows them about all the time. If one can make a cage of wire netting it is probably the best course. I started by trying to keep my piles anchored with pieces of wire netting, but it was not very satisfactory. Next I dug a pit and put them into that and had wire netting over the top. This meant that the contents of the pit had to be moved each year to allow the new lot of leaves to go underneath. I don't think my home-made leaf-mould ever became thoroughly rotted, but that is not necessarily a bad thing in heavy soil. The worms soon reduce it to fine particles, aerating the soil at the same time. I know one good gardener who makes no attempt to rot down the leaves from his trees, but merely sweeps them on to the flower beds and leaves the rest to the worms. The idea is a good one but there is no way of stopping some of the leaves being blown back on to the grass and paths and the result is an untidy garden for a few weeks. Also I should be afraid of encouraging slugs—of which I already have more than enough—as there is nothing they like better than to hide under a thick layer of wet leaves. So I still sweep up the leaves from the big sycamore on the lawn and it is a job that has to be done every day for a few weeks after the leaves begin to fall. There are not enough to stack separately, and I would not want sycamore leaf-mould anyway, so now we put them on to the compost heap with the other green material.

If I had border-line trees and shrubs to protect for the winter I should use them first for that and put them on to the compost heap in the spring. Leaves make a pleasant way of protecting subjects that might be damaged by frost and are far less unsightly than covering the plants with polythene. Perhaps it makes a little more work but anything is better than having the garden full of polythene bundles in the winter. To do it the plant is encircled with wire-netting held up with canes. Another circle of wire netting is fixed about 1 ft outside the first and the cavity filled with leaves. If one prefers a green rather than brown effect trimmings from conifers or other evergreen trees can be used instead.

As I have always planted everything in the garden with a mixture of peat and coarse sand, this must have helped break down the soil, but clay is very tenacious and tons of sand and peat would be needed to make it completely friable. At least my top soil is no longer sticky, and if worked after rain is easy, but it still gets hard in dry weather although there is plenty of moisture lower down. It amazes me how well plants in clay stand up to drought, except those, such as phlox, that have their roots near the surface. Soil in which seeds have been raised is always returned to the garden and given to small plants that need a little encouragement. I now have a pile of turves, stacked upside down, to provide soil for sifting for potting mixtures. Before that the orchard had to be raided, and I found that the soil just under the turf was very good for the purpose. Of course anyone who has a plague of moles has a source of ready prepared soil, but no one wants moles even for fine soil.

If plants are replanted in the soil from which they are taken they settle down more quickly and soon 'take' to their new medium. For this reason I buy as many plants as I can from a local nursery which has the same limy clay that I have. Plants I buy from another nursery, which has dark peaty soil, often will not settle down in my garden, however much I cosset them. The texture of soil is very important and the more lightening material that can be incorporated in clay, the easier it becomes to introduce plants from other types of soil.

DRAINAGE

Drainage is more important on clay soil than on any other. Light sandy soils, for instance, drain so quickly that very little moisture is retained unless a great deal of humus is added. Clay soil retains moisture, and plants do better in it in drought than in lighter soil, yet very few plants will grow with their roots in water. As the deeper one goes the more impenetrable the clay, it is obvious that lower down the moisture simply cannot get away and in very heavy soil it is sometimes necessary to arrange some kind of drainage. In a garden near me made on a steep hill, the ground was terraced and I suspect the part which was made into a rose garden was mainly subsoil. Large holes were dug for the roses and they were planted with the utmost care, but they did not prosper. Then it was discovered that the roses were literally standing in water. It was not difficult to drain it away by making a series of trenches and because of the steepness of the site it was necessary only to divert it to the edge of each terrace. If this had not been possible the channels would have had to lead to a sump (or pit) with stones in the bottom.

For plants such as pinks and silver-foliaged varieties I make raised beds by building small walls about 1 ft or 18 in high. The top soil is carefully removed and the subsoil broken up with a fork. On this goes a layer of stones, crocks or broken bricks and then the top soil is put back. For a few rare specimens the bed is finished with a mixture of coarse sand, peat and loam, but usually top soil is enough with sand and peat for planting. Much of my garden is on different levels which means there is good drainage. Retaining walls are better constructed without concrete but if the ground is supported by a concreted wall, holes for drainage must always be left in the wall.

The problem of drainage does not arise in rock gardens which are made by building up with stones, and are usually filled with a planting mixture in which plenty of coarse sand is used. The use of chippings on the surface stops the soil from caking and preserves the moisture. Moisture-loving plants grow well at ground level, eg mimulus, small ferns, white or double caltha (marsh marigold) and *Sedum pulchellum* all do well in marshy conditions.

Having many walls I have in several cases made rock beds against them and here plants do well because the drainage is good. I have quite a number of raised beds against my outbuildings, which enables me to plant climbers against the buildings. This would not otherwise be possible because there is concrete against the walls. These beds are also ideal for small bulbs which need very good drainage as they always have a thick layer of stones and crocks before the planting mixture is added. I have noticed that plants in them do far better than those at the foot of the wall, even if there is no concrete. These low-lying beds often have poor drainage and some plants do not do at all well. However, hellebores and hostas do not seem to mind a heavy, badly drained soil, and some primroses do well in such conditions, even the difficult doubles.

A lawn on clay soil needs good drainage more than any other part of the garden. When we made our garden the lawn was prepared with great care, levels were exact and good soil was used for the top layer, but nothing was done about drainage. It has always been difficult to keep in good condition, with lush vegetation in one place and sparse growth in another, and moss and weeds all over. My husband worked with our garden boy to make this lawn and I do not think he realised that clay needed special treatment any more than I did. The soil in his garden on Sydenham Hill was the light rich loam of old gardens that have been well tended and manured for many years. He certainly realised the difference in the soil after he had been at Lambrook some time and gave me lyrical descriptions of the dark friable soil that had the consistency of grated breadcrumbs and never caked. Clay of course cakes at the least provocation and is particularly prone to become compacted when badly drained. If one walks on it in wet weather it becomes as hard as stone when there is a drought.

I know that to get a reasonable lawn I should put in drainage, and if I were younger I should do it instead of other improvements in the garden which to me are more exciting. As the ground has a slight slope to the house the procedure would be to dig a diagonal trench across the lawn and put drainage pipes in the bottom of it. These would be covered with a hard core of broken bricks, large stones, clinkers or cinders, with a layer of gravel or small pebbles on top. Side trenches dug in herring-bone fashion would be made to slope slightly to the main drain, which would lead to a deep sump, filled with hardcore, gravel and then, of course, with top soil, and levelled up with the other trenches and the lawn. I could also improve the lawn by levelling places which have sunk during the years, which would mean removing the turf and adding top soil. In other places I think returfing should be done where, when the lawn was made, the dreaded subsoil got to the top.

The other so called 'lawns' in the garden give far less

trouble. The small piece of grass in the middle of the barton is on a slope and so drainage is simple. This was an afterthought and I planted it with tufts of fine grass I found in the flower beds and in the road. Other pieces of grass which are reclaimed orchard are far better than the large lawn we made on part of the barton. The orchard grass was coarse to begin with but years of mowing have brought it to a reasonable fineness, and of course one has the advantage of excellent top soil below the turf, rich in bacteria and humus. Whenever we need good soil we go into the orchard, remove a yard or two of turf and take the soil we need from the top few inches. Down below, of course, there is clay but the soil immediately below the turf is infinitely better and easier to handle than that in the garden. Where we have taken in parts of the orchard to make the garden bigger, everything planted does exceptionally well and no compost is needed to make the soil manageable.

COMPOSTING

In these days I think every gardener is converted to compost but not everyone has the same idea on how it should be made. It depends, of course, on the size of the garden and the time the gardener can spare for the job.

I make a lot of compost because in my thickly planted garden a great deal is taken from the soil and so I try to return it, and a little more, in this way. It is not difficult for me as I have plenty of room for the two big heaps required for a two-year programme and plenty of material with which to make it. The more compost used, the more luxuriant the growth, and I am lucky to have the services of a good man who makes my heaps for me and wheels the finished product to every part of the garden where he dumps it in piles for me to spread carefully among the plants.

I am sure that the bigger the heap the better the compost will be. Recently I have read in several books that a compost heap should never be bigger than a certain size (I did not make a note of the exact measurements), and it was also stated that it is better to have several small heaps instead of one big one. Where this idea originated I do not know. Of course in a small garden there are limiting factors both in regard to space and to material, but the heap should be as big as possible because the principle of making compost is to raise the temperature as high as possible so that fermentation takes place, which causes the contents to decay and allows the bacteria to do their work. A heap that is too narrow soon dries out and the whole process is slowed down: 5 ft is the minimum width that should be used, and anything between that and 12 ft is good, but if the heap is too wide air will not easily get to the middle. A good size for an average small garden is 8 ft x 12 ft but there is no limit to the length that the heaps can be. In a small garden the compost is usually wanted quickly and this can be achieved by turning it over periodically.

In no part of the garden should there be untidiness and compost heaps should be neat and never unsightly—again the size of the garden determines the method to be used. My first compost heap was enclosed by neat hedges of *Lonicera nitida* which had to be clipped. I made it on a concrete base which is a system to be condemned as it is important to encourage earth-worms and I did not get many. It was made at the top of a bank, and the base sloped very slightly towards the bank. I had gutters made round the heap so that all the valuable moisture from the compost drained into cans set in niches in the bank below. This source of liquid manure was useful but not as important as the earth-worms, so I changed my system and the site.

All green material from the garden should be left to dry and wither before being used, so I decided to make two heaps side by side, the one on the left to be the finished heap and that on the right to take the current material as it is collected. The heaps are 18 ft x 8 ft each which gives enough room to spread out the new material all over the heap for it to dry off before more is added.

A permanent site for compost heaps is recommended. The soil then becomes richer every year with bacteria and each new heap has a good start. Compost heaps should be made on earth and not turf, and my first double heaps were enclosed by walls made by the up-turned turves. This meant trimming the turves, which soon produced a fine crop of grass whiskers, and looked most untidy.

My next heaps were therefore enclosed by bales of straw, two bales being a good height, and these were placed to overlap as if they were bricks. Stout posts were driven into the ground at the corners and the bales kept in place by two lines of strong wire fixed to the posts. I used straw because I thought it would hasten the heating up process, and the old straw could be used in making the next heap. It worked very well in normal years, but after the endless rain of 1968 the straw bales began to disintegrate earlier than they normally do and had gone to pieces by the end of September. This did not affect the compost as by then the heaps were more or less firmly formed, but it looked untidy and we had to be continually sweeping up the paths round the heaps. I shall not discontinue a system that has worked well for seven or eight seasons but it proves that straw bales are not a good idea in places where there is a constantly high rainfall. Strong wire netting fixed to stout posts would be satisfactory, and the compost could be contained by wood, or even hurdles, but please not corrugated iron. Corrugated asbestos is not unsightly but would cost more.

The compost is spread in November or December and the new heaps begun. I try to cover every part of the garden every year, and give particularly generous mulchings to beds under walls and under trees, and to such plants as roses and paeonies. When we have carted away the compost the new frame work is started, with the enclosure for the new heap being made first. There is usually some of the rotting straw from the old heaps left and to get rid of it we put that at the bottom of the new heap, then 6 in of the garden waste that has been building up during the year. The next layer in my case is 2—3 in of manure as I live in a farming community and the farmer who rents my orchard for grazing is willing to pay part of his rent in farmyard manure from his herd of cows. Gardeners not so lucky might be able to get fish manure or other animal waste, but if no organic material is available a vegetable activator is the next best thing.

The layer of manure is covered with a thin layer of good soil mixed with wood ash or agricultural lime. On our limy soil I use wood ash, in fact the ash from the bonfire is often used without soil as we burn a great deal of soil with weeds. I once used my natural clay soil but it remained in the same state as when it was put in the heap, and after a year looked like a layer of almond paste in a cake. Soil for this purpose must be good top soil and it should be used if available as it contains bacteria which speeds up the decomposition processes.

These layers are repeated with a generous layer of manure at the top before the last layer of earth and wood ash. The top of the heap is covered with loose hay to keep in the moisture and heat. It is sometimes recommended to make the centre of the heap a little lower than the outside so that all the rain that falls gets into the heap. One advantage of making compost inside 'walls' is that the top remains flat. When we made our heap in the hedged enclosure I could never get the material spread out properly and the finished product was more like a Noah's Ark than a compost heap and rain just ran off it.

Loose packing means that a certain amount of air gets into the heap, but that is not enough. Some gardeners thrust a crowbar into it at intervals and work it round to make holes 5 in or 6 in in diameter. Another way is to stand upright pipes at intervals down the middle of the heap and remove them when the heap is made. I use short pieces of pipe which are raised as the heap is made and are left sticking out of the top to keep the space open.

When the second enclosure is made we start again to collect material for it. I have read that as many different types of material as possible should be used as each plant has something valuable to offer. Most experts recommend that weeds should be used as well as cultivated plants because the extreme heat should kill all weed seed but I never put in weeds, evergreen leaves or really woody stems from shrubs. I used to add annual weeds to the heap but there is no question that since I have used so much compost I have had many more weeds, although not necessarily from the compost of course. One can argue that in a garden that is surrounded by fields, orchards, grass verges and mixed hedgerows the air is naturally full of weed seeds. Opposite my garden there is a piece of land belonging to the village inn that is a solid mass of dandelions; the countryman usually leaves his vegetable patch untended until it is time to dig it for the new season's sowing, and having improved the fertility of my soil I am at the mercy of every seed that blows. Gardeners who do not 'hold' with compost, in other words cannot be bothered to make it, argue that its use encourages weeds in the garden, but while we have wind to blow the seed about and rain to make it germinate we shall have weeds. The ones that really frighten me are not the honest docks and dandelions and sow thistles but the little creeping horrors such as *Veronica filiformis* and *Helxine soleirolii* which propagate themselves by leaf and stem, for every tiny scrap left in the soil will grow. Now I burn all weeds, to be on the safe side, and having barrows of different shapes, one is automatically used for compost material and the others for stuff to be burnt. In a garden so thickly planted as mine we get plenty of material to compost: grass mowings are spread thinly over the heap so that they do not form a soggy mass, and nearly every day there is a bucket of 'gash' from the kitchen, vegetable refuse, tea and tea-leaves, coffee grounds and crushed egg shells. The liquid is useful in the heap, which should be watered in dry weather. This is more important when making a small amount of compost quickly than when it is left for two years before being used, for in that time there is enough rain to make it moist enough.

Where space is limited and compost is more urgently needed the process can be speeded up considerably. The same ingredients are required and it is still best to let the green material wither before adding it to the heap. For quick results watering in dry weather is necessary, and the heap should be turned three weeks after it is made so that the compost on the outside goes to the middle and gets the advantage of heat and the greatest bacterial activity. A five-pronged pitchfork is the best tool to use. A second turning is done five weeks after the first one and the compost is then ready in about three months. In a very cold climate it may take longer. It should not be used until really decayed but it can be left after it is finished for some months, in which case it should be covered with a thick mulch or canvas.

Containing a small heap is not easy. Some people use

used it can be firmed by foot in the normal way, and then watered.

When plants and shrubs are started off with peat and sand they settle down at once and continue growing well when their roots come to clay. When possible I buy plants from a nursery on soil similar to my own because I have found that, after being reared in one type of soil, some plants resent very much being transferred to another, especially when the new home is on clay. It would not matter if it were the other way round. I have had to give up buying plants from a famous Highland nursery, which has acid peaty soil, for this reason. I do not expect the lime-hating gentians to take to my limy clay but I did not think I should have difficulty with such an easy plant as *Geranium sanguineum* (in a rather special colour break, of course) or *Clematis tangutica,* which grows well for me in its ordinary form. The one I bought was an improved version, with larger flowers and seedheads which looked as though they had been brushed into miniature toques. Both these plants died. Ordinary plants which I buy from another nursery on rich peaty soil do very well but anything that is at all temperamental usually refuses to co-operate.

Plants with long tap roots do not always settle down as they should; I suppose the roots come to a clay subsoil and get discouraged. When replanting such a species I usually reduce the length of the root with a sharp knife. Tap-rooted plants do not last long in clay: I used to grow lupins but they were seldom any good the second year, so I realised I should have to treat them as biennials and raise them afresh each year, but as I have not time for anything but perennial plants, I had to give them up. I was particularly sad when a lemon-yellow tree lupin died because I thought that might be more amenable than the herbaceous ones.

Plants that are inclined to rot off at the crown need more

careful treatment when grown in clay than in other soils. All the members of the dianthus family need special care and if it is not possible to grow them on raised beds, a few chippings round the crown of the plant will often save its life. With rare primroses I use peat, not only to prevent them rotting but to cover the young roots which often appear almost on top of the soil. The petiolares primulas are not easy to grow in southern counties although they thrive in Scotland, where conditions are more like their own in nature. In a Devonshire garden the beautiful blue *P. bhutanica (P. whitei)* was grown successfully for many years with peat packed round the crowns to prevent rotting from winter rain.

A Cotswold garden I know has soil that is not clay but very stiff and by no means easy. Special plants such as orchids are put in a lightly shaded position and the soil is covered thickly with moss peat. Few terrestrial orchids take kindly to cultivation but the collection treated in this way is growing well.

Whenever possible rainwater should be used for watering. Containers for this should be open to the air and only covered with wire netting to stop leaves or birds getting into the water. Large wine casks painted green are not unsightly against buildings which are guttered, and wire netting can be fastened to the back of the cask with staples and stretched over nails in front.

WHAT NOT TO GROW

Before I discuss the plants that I have found grow particularly well on my clay soil I will mention some of the failures. I cannot, of course, be absolutely certain that it is the clay to which they object but as I know they are all lime lovers I feel certain it is the structure of the soil not its chemical construction that is the trouble.

The first plant about which I had serious suspicions was

Scabiosa caucasica. I have had both the lilac-blue 'Clive Greaves' and the lovely white 'Miss Willmott' but not for many years. At the beginning they seemed to do quite well and I remember describing the delicate and perfectly symmetrical detail of the flowers, with their touches of white and green, but they dwindled and finally rotted away at the crown, and after a few more attempts I decided not to try any more.

The same thing happened with pyrethrums but they didn't last even as long as the scabious. My first ones came from a garden in Buckinghamshire where the soil is gravel and I remember my hostess digging up her big clumps and giving me half of each, but they did not stay with me long. It must have been in the early days that I had the scabious as I remember I had time for flower arranging and spent whole days running in and out to find just the bloom I needed to complete my masterpieces, to the great annoyance of my husband who thought I could make do with the material I cut in the first place. Scabious helped me with delphiniums and I was delighted to find how its lilac flowers brought warmth into the scheme of blue, which is a cold colour.

Adonis are charming little spring-flowering plants I have been trying to grow ever since I started gardening. I have had them in many shady positions in the garden but they just pine away. I know they enjoy a limy soil so I am sure it is the clay they dislike. Any plant that flowers in March is something to look for and I tried hard to please *Adonis amurensis,* with their rounded flowers of gleaming gold nestling in little ferny ruffs. The foliage is fernlike too and it grows rather coarse after the flowers are over. I remember my excitement when I managed to get the double form of this fascinating little plant and I did everything I could think of to induce it to settle down but it went the way of the scabious, as did *Adonis vernalis,* which has flowers like many-rayed sunflowers amid the ferny foliage. Although to me adonis are typical woodland plants they seem to do best in full sun.

The tall thalictrums all do extremely well with me, seeding themselves in a mild way, but the attractive dwarf *T. diffusiformis*, a relative of the tall *T. dipterocarpum*, disappeared in its first season. It grows to about 1 ft and has comparatively large mauve flowers and typical fernlike foliage. I put it in a semi-shady corner, for I think that *T. dipterocarpum* does best with a little shade, and the crown of the plants should be about 1 in below the surface.

It is only after thirty years that I have managed to keep the entrancing 'balloon flower', *Platycodon grandiflorus*. Always before it has done quite well in the first year but did not reappear in the next. This member of the campanula family is more attractive in the bud stage than when it opens. The petals have a dull finish and are joined together to make a flattened 'balloon' and then open to large saucers in varying shades of blue. *P.g.* 'Apoyama' has large flowers in rich purple-blue, and in *P.g.* 'Mariesii' the flowers are also big but deep blue. Both are dwarfer than the species itself. That the most recent plants have survived two winters makes me think that in at least one place in the garden I have managed to break up the clay sufficiently for them.

It seems strange that I cannot keep pulsatillas *(Pulsatilla vulgaris)* in my garden. I have all the lime they want and they do marvellously on chalk, but, although they last for a short time, they soon dwindle and eventually disappear. In a friend's garden about twenty miles away which, I should have thought, has the same soil, they increase and seed themselves as they do in the wild. They are, I think, plants that dislike disturbance, so I left mine completely to their own devices as my friend does. Probably the reason why my pulsatillas died was because they found the soil too moist. They like firm, hot soil and

Somerset clay does not always live up to these requirements. They probably did well in my friend's garden because the large area where they grew and multiplied was paved, and though I grew some of mine in paving too, they had larger pockets where the clay became wet and squelchy. I know several places where pulsatilla still grows freely and they are all on chalk downland.

There seems to be some confusion between chalk and any other soil which has a lime content. I am often asked if a certain plant will grow on chalk, when I am sure that the enquirer really wants to know if it is a lime hater. Chalk has a different construction and is always limy whereas clay can be acid or calcareous. The late Sir Frederick Stern made a magnificent garden on chalk, which is the subject of his book *A Chalk Garden,* but he broke the chalk up when he was planting.

I often wonder why the cottage chrysanthemums do so well in cottage gardens and yet I have a hard time to keep them. Many gardeners may not know these charming simple flowers because there are today so many supposedly more attractive varieties. To me chrysanthemums belong to the autumn, their scent goes with the smell of bonfires and wet leaves, and I do not want to see them blooming in the garden in July. I used to get enthusiastic about Korean chrysanthemums until I discovered that if I wanted to grow them I should have to buy new plants each year. Of course they are not completely hardy, but in this part of England I do not think that was the trouble, and my subsequent experience proves it.

These dwarf chrysanthemums with the small double flowers in faded colours have always had a fascination for me. The brightest of them, 'Jante Wells' has—or shall I say had, for I never see it now—bright yellow flowers, and even when the others were doing well I could never keep it, although it was a reliable perennial in some gardens.

It must be twenty years ago or longer when I first had 'Dr Tom Parr', with flowers of faded rose madder. At first there were big clumps in different parts of the garden but they dwindled over the years and now it has gone altogether. The dull magenta-pink 'Anastasia' increased so much that I didn't know what to do with it. I haven't got it now, yet I know cottage gardens that are full of it; in fact it is the only plant visible in some of them. Friends who had a garden in London on light soil gave me the pale pink form of 'Anastasia' several times, and I also had it in bronze, but not for many years. Four varieties that are still going well are 'Emperor of China', which is also known as 'Cottage Pink', 'Cottage Bronze', a dark crimson single with a yellow eye, which my father used to grow, and a soft, pale pink single called 'Innocence'. The first three are tall and it is difficult to grow their lanky stems with any degree of grace. Some I tie up to a wall, and the bronze I planted under a shrub, *Helichrysum gunnii*, so that the chrysanthemum could grow through its stiff dark stems. 'Innocence' is comparatively dwarf, usually from $2\frac{1}{2}$—3 ft, and grows quite well if supported by a semi-circular iron stake, so long as it is put in early enough.

I have now acquired a new cottage chrysanthemum, 'Cottage Yellow', which makes a brave show in a council house garden in this village, but it may not do the same thing for me. The garden where I see it is high above the road, and the chrysanthemum is the only thing in it that flowers at a certain time of the year, so there is no competition. But it will find competition in my garden, although I leave its immediate surroundings clear so that I can see the clean Chinese yellow of its flowers.

Plants with rather spiky crowns are inclined to rot at ground level. I have tried several times to grow *Puya alpestris*, with its lurid flowers that have a faint resemblance to a bird of prey.

and over the door to give leafy protection to the clematis stem.

Nearly all the different clematis I have grown have done well in my limy clay. Some are recognised as more temperamental than others. When nurseries describe the variety as 'not vigorous' it usually means they are somewhat temperamental. 'Mrs Oud', which has large milk-white flowers with dark centres, can be unco-operative and to grow it successfully is a horticultural triumph. I have never tried to grow this variety but I have struggled for years with *C. florida bicolor* (syn. *sieboldii*). It needs a sheltered wall and my plants have stalwarts on each side as well to keep out draughts; I am sure they do better growing on another plant than on man-made supports. I have given one the companionship of the more ambitious *C. macropetala* and it seems to be doing well. I have never tried the white form of *C. florida* because I think it is mainly the exquisite colouring that leads us on to buy one after another of the true plant in the hope that one will succeed. The petals are very pale green (not white as some of the catalogues say), with a domed central boss of bright violet, surrounded by petaloid stamens, so why struggle with white flowers when there are other white-flowered clematis far easier to please?

I think most of us try to coerce our clematis too much and make them grow just how we want and not as freely as they do in nature. Most of us have to grow clematis on walls and even with adequate wires or areas of wire netting a great deal of training is needed at the beginning of the season if the plant is to be spread out to cover a wide area. Pinching out the tops of the growing stems periodically makes them grow side shoots and prevents the long stems from muddling together instead of spreading out fan-wise.

Left to themselves clematis supposedly growing against a

wall will find other positions or new companions they like better. I have a deep blue *C.* x *jackmanii* 'Superba' trained over the wall near my garden gate. Always it has decorously made its way to the top of the wall and then proceeded to clothe the other side, making use of the wire netting provided for the purpose. This year it managed to get into a bird cherry growing nearby and festooned the pale green branches with sheets of deep purple flowers. I hope it will go on doing it.

For several years now a strong plant of 'Ville de Lyons', with rich red flowers, has only partially covered the wall against which it is trained and many stems prefer to ramble among the small shrubs and lavenders growing below. It is always too late to do anything when one discovers what is happening for the stems of clematis are very brittle and any attempt to untangle them and get them back where they belong usually results in breaking most of them. The only way to train clematis is when the shoots are very young and it is a daily job to sort them out and start them in the various directions they should take. When there are as many clematis plants in the garden as I have I should do nothing else if they were all to be trained properly, and the more I work with them the more convinced I am that they should be grown in the garden in the way they grew in nature if this is possible. Give them another plant to grow through instead of using such artificial props as wire netting or trellis and they do the rest themselves. This is easier with the species than some of the big-flowered types, although roses and clematis go well together, and the large flowers make a show on fairly large trees, such as apples and pears. The clematis should be planted about 1 ft away from the tree and a certain amount of trouble is needed to find a place on the north side of the tree between roots where a big enough hole can be made. The clematis should still be tied to its bamboo and this is tilted to the tree and secured to it. As the clematis grows it will need to be tied to the tree but when it reaches the branches it will take care of itself and make a far more natural display than can be achieved by man.

The small-flowered species show up well on shrubs that flower at a different time of year. Each year I find a few more plants that will make good hosts and a few more attractive clematis to grow through them. The different hybrids of *Clematis texensis* are particularly good for this as the flower buds occur at the leaf axils and need to emerge into full sunlight to open. One of my favourites is 'Duchess of Albany', which has the usual *texensis* urn-shaped flowers in a warm pink with a cherry-red central stripe on the outside of each petal. I used to grow this clematis over a broken wall but find it happier weaving its way through a bush. In one place in the garden I have it growing through *Osmanthus delavayi* and in another it has several low-growing shrubs in which to ramble. It seems to favour most a large mound of purple salvia, *S. officinalis* 'Purpurascens', and the upright carmine urns show up well against the soft grey-purple foliage. So far none of the trails which have found their way to *Choisya ternata* have flowered. I don't think this shrub, with its light green glossy foliage, will make such a gentle setting for the clematis.

The flowers of 'Gravetye Beauty' are a deeper shade of pink and though they start as ruby bells, they open to star-shaped flowers slightly paler in colour. I have given it a dogwood, *Cornus controversa* 'Variegata', as a host plant. This shrub has a very distinct manner of growth, with completely horizontal branches growing from the main stem. They make tiers when the shrub is big enough but my plant appears to have stopped growing up and the main side branch is now a wide canopy covered with delicately variegated white and green leaves. It is just above eye-level and I think the clematis flowers will show up very well.

A taller shrub is needed for the small flowers of 'Etoile Rose'. These are campanulate and the four petals are soft carmine in the centre, edged with silver. Instead of growing upright from the stems they have a nodding habit and would look lovely growing through other plants on an arch. The hebe I chose is not nearly big enough and I am wondering how a small Judas tree *(Cercis siliquastrum)* would do. Its flowering is over long before the clematis starts its autumn season but the large round leaves might hide some of the flowers. Something more flimsy such as the weeping alder or weeping birch might be better. The clematis growing through the weeping alder in my garden was a self-invited guest, and makes a good example of the superiority of nature, as a planter compared with human effort. I first had a *Clematis viticella* on the alder, but the small purple flowers hardly showed among the tangle of dark stems and flowers. One of the paler forms of *viticella* such as 'Huldine', 'Alba Luxurians' or 'Little Nell' would have been splendid but the flowers of the type plant were too sombre. I transferred *C. viticella* to a large bush of variegated cornus and had not found a replacement for the welcoming arms of the alder when I saw that nature had got there first. *Clematis flammula,* seedlings of which have appeared in several parts of the garden, had sown itself at the foot of the alder and was festooning the top part of the tree with clouds of scented cream flowers.

Another seedling appeared near the lovely single pink rose 'Cupid', which is trained to a pillar. I disentangled the muddled mass and hooked some of the stems on to the leaf stems of the rose; very soon it was all round the pole, making a cloud of cream around the rose and filling the air with strong perfume. The rose, of course, is one of those pleasant

varieties which produce large colourful hips after flowering, as large as small tomatoes. They look singularly attractive glowing through the filmy flowers.

One of the most beautiful of the late-flowering types is 'Alba Luxurians', which makes a picture for many weeks. My first plant was on an east wall and it had to compete with rather big shrubs. To begin with it made a good show but when the silver-variegated privet, *Ligustrum ovalifolium* 'Variegatum', got big, and *Rosa moyesii* even bigger, the clematis seemed to lose heart, and though it still flowers each year it does not show up well. Luckily I had put another on the north-facing wall at right angles to the front of the malt house where it has little competition. Here it is a wonderful sight for weeks; I keep thinking it is coming to an end and then another part of the climber comes into bloom and there is another patch of fluttering white flowers, each one slightly different in shape and with varying-sized touches of green on each of the four sepals. The dark centres of the flowers give the plant great character. An 'Albertine' rose growing on the south side covers the top of the wall and at one end there is the red-flowered clematis 'Mme Jules Correvon' which flowers from June to September. So I think 'Alba Luxurians' is not so much a clematis for growing with other plants as for exploring the possibilities of a wall or a roof.

The winter flowering *C. calycina (C. balearica)* sowed itself under the same wall but there is not room for its tremendous growth and I dug it out. The mass of dark ferny foliage that this plant makes would be useful on some large bare wall or to cover up an eyesore but it has the reputation of not being free-flowering and it has lived up to that in my garden. The parent plant festooned everything in sight, including a strong bush of the rose 'Zephyrine Drouhin', a fuchsia and *Chimonanthus praecox* var. *luteus*, but gave me few of its small green flowers, freckled with red inside. On the credit side is the fact that it is evergreen and its few flowers are produced in the winter. I know gardens where it blooms with real abandon, without making nearly so much growth, so it may be that there are different forms, or I feed it too well.

4

OTHER PLANTS THAT DO WELL ON CLAY

Recently there was an article in one of the gardening magazines by a well-known gardening writer entitled 'Plants for Impossible Clay Soil', in which he painted a gloomy picture of the difficulty of growing anything in bad clay soil. But he did concede that there were some families of plants that do well in clay.

One genus he mentioned was the campanula, and I agree that most members of this family do extremely well in my garden. One of the first plants I put in was the old-fashioned *C. latiloba.* It is a real cottage-garden plant, with most excellent foliage that can make good ground cover. The saucer-shaped flowers are white or blue and they grow tight against the tall stems. I think it would grow anywhere; it is indifferent to position and will cover the arid soil under holly trees. Even in its improved form, 'Highcliffe' (or 'Percy Piper'), it is too rampant for most people, but I find it useful for growing under trees, including conifers, and enjoy its rather pale green foliage in winter. It is sometimes called *C. grandis.*

Another variety that does well on clay is *C. lactiflora,* which has a pleasant habit of seeding. The type plant is the soft milky-blue that the name suggests and it seems to attract attention more than any other campanula. The large heads of small flowers mix well with summer plants and are not out of place in a wild garden. Its seedlings can be white or shades of blue. The pale pink 'Loddon Anna' is particularly charming, and there are two dark blue forms, 'Superba' and 'Prichard's Variety'. I often wonder if the dwarf 'Pouffe' was a natural break or a man-made plant. It is most useful for small beds or in front of taller plants, is easily increased by division and has never with me exceeded its normal 9 in.

Of the two types with large hanging bells I like *C.* x *burghaltii* better than the various forms of *C. latifolia* merely because it flowers throughout the summer and almost into winter. Its slaty blue flowers are delicately shaded but they always seem a little big for the plant, whereas *C. latifolia* carries its large hanging bells on strong 4-ft stems. But its season is short and something has to be grown in front of the plant to hide the gap.

I don't believe *C. glomerata* is a nuisance in all gardens; it likes my clay so much that it runs in all directions and once planted in a bed will quickly cover it. The named varieties are not so invasive and their clustered heads are a deep violet-blue. 'Purple Pixie' is another rather neat growing form with 18-in stems and deep violet flowers. It is a great disappointment that I have not been able to do much with the white form of *C. glomerata*. I know that sometimes the white versions of plants are not as robust as the coloured forms, but I should have thought that an irrepressible plant like this one, *C. glomerata,* would have had a reasonably strong white counterpart.

The most graceful campanulas for growing in flower beds are the various forms of *C. persicifolia,* the peach-leaved bellflower. The ordinary variety in blue or white seeds about the garden in a modest way, its neat clumps of evergreen foliage usually fitting in quite well in the position it has picked out. There are named varieties of this, 'Telham Beauty' which has large blue open flowers and the white 'Snowdrift', but I am quite happy with the rather smaller type plants which nature bestows on me. The best of the double whites is 'Fleur de Neige', but the one I grow is not quite so big and is I think the older 'Moerheimii'. I have tried 'Fleur de Neige' but it did not settle down in my garden and after a year or so gave up the struggle. The double white I grow now is much easier, it increases well and flowers for a long time. 'Pride of Exmouth' is a blue double *C. persicifolia.* It has not the appeal of the white forms although the soft lavender-blue of its flowers makes it easy to place in the border, but none of the doubles compare in gracefulness with the old-fashioned cup-and-saucer types, and of these the white is again more beautiful than the blue. For one thing it sometimes has touches of green on the tips of the cup and petals, and the plant is more generous with its flowers. I moved heaven and earth to find my blue cup-and-saucer campanula and I would not like to lose it, but there are some flowers that are more endearing in white.

This is the case when *C. poscharskyana* is considered. Sometimes I wish I had never introduced this rampant campanula to the garden because it can become a real weed in places. Put it on a wall, where its long trails of light blue flowers can hang down, and it receives nothing but admiration, but if the wall is a supporting one and there is a bed behind, it will encroach further and further and its persistent white roots will find their way into the choicest plants. I expect the white form would in time be as troublesome but it has never ventured beyond its wall so far and I find the flowers, which are white with a pale blue wash, far more beautiful than the blue ones. The other easy rock campanula, *C. portenschlagiana* is not quite so invasive as the other and because it is smaller and neater it does not become a menace in the garden. Both these plants flower on and off all through the summer, and the smaller one carries on well into the autumn. So if your soil is clay in its worst form, bare places can be made beautiful with these willing campanulas.

The plants we call Japanese anemones (*A. japonica* or more correctly *A.* x *hybrida*) do well in clay. Even a scrap of root left in the soil will cling to life and eventually become a flowering plant. They come up between stones and at the bottom of walls, and once in a garden I defy anyone to get rid of them completely but, like many plants difficult to eradicate, they take time to settle down and nurseries usually sell them as small pot plants.

I still think the tall white *A.* x *hybrida* 'Louise Uhink' is the most beautiful anemone of all. I like to see it massed against a dark background and in one place in my garden a good planting has a curtain of *Clematis rehderiana* on a wall behind it. I grow many pink and ruby-coloured anemones and these have more appeal to my visitors than they do to me. Some are almost double but none has the dignity and character of the elderly Louise.

If the Japanese anemones are good standbys the smaller pink *A. hupehensis* might almost be called a coloniser. This is the plant that the late Hon Robert James allowed to wander through his borders in his garden St Nicholas in Richmond, Yorkshire. The flowers are small with a lilac tinge and dark reverses to the petals, but they last a long time and are less stately than the others. A single plant I put in a shady bed has now filled it completely and is making leisurely progress into the bed on the other side of the path. I don't want a jungle in that particular part of the garden and from time to time remove the advance contingent, but it makes little difference for very soon the ground is as well covered as before.

when we know them we find they are quite easy to grow—and to know. I had an old cousin who always told me I was gardening above my station when I introduced something a little unusual to the garden and I am sure she was not really sorry when I didn't succeed.

Saxifraga fortunei is a woodland plant that has a reputation for being a little difficult and the idea has grown up that it mustn't be disturbed or divided. I find it one of the easiest things to please if I plant it in the lower levels of my ditch garden where the heavy soil is usually moist and there is shade from willows and a quince tree. I divide it quite often and every year in late autumn the plants are a mass of fluttering white flowers. The variety with richer coloured leaves grows in a less shady spot because in 'Wada's Variety' the bronze leaves are lined with rich crimson and glow like stained glass when they catch the glint of the sun. The flowers I do not think are quite so good as those of the ordinary type.

Another plant that is sometimes discussed with marked deference is *Iris innominata*. It must have neutral soil, must grow in full sun and should never be disturbed. Mine have none of these things and yet they flower well and increase. I did try my plants first in full sun, according to instructions, but they did not do so well as when I moved them to a shady spot in very heavy soil. In addition I know they are often subjected to trampling as visitors will use their corner as a short cut to a collection of variegated plants in a bed behind. The flowers of this iris can be tangerine yellow with falls netted with chocolate lines or a very gentle shade of orchid pink. The foliage is narrow and stiff and as it grows almost horizontally it makes a good background for the flowers.

The pineapple lily, *eucomis*, which comes from the Cape, is sometimes treated as a greenhouse plant but I know gardens near London where it grows outside, so I think it is reasonably hardy but I always give it a south position. It has large fleshy leaves, and tall spikes of glistening green flowers growing against the fleshy stems, with a tuft of leaves at the top like a pineapple. The species most commonly seen is *E. bicolor*, which is green with purplish edges. *E. punctata* is green and brown and *E. regia* green and purple. It is usually recommended to grow this plant in light, rich soil, but they do so well in my clay that I have difficulty in supporting the stems. They are so thick and heavy that they snap in a strong wind, although supported. In very exposed districts they can be covered with ashes or decayed manure in winter.

Alstroemerias, which are commonly known as Peruvian lilies, do extremely well when they get into their stride. I made several attempts before I managed to get them to grow, but that was through ignorance. I treated them like any other plant, not realising that they have roots as brittle as glass and need the greatest care if they are not to be damaged. When I discovered that it was possible to buy small plants in peat pots, so that pot and all could be planted, the rest was easy. The ones to buy, of course, are the Ligtu hybrids, which come in shades of flame, buff and champagne. I was lucky for the two groups I planted in south-facing beds happened to be in two separate colours. In one place they are that pale shade we call 'champagne', and the other lot are all flame. Usually there will be all shades in one planting. Once established they stay for ever. Gardeners who are unfortunate enough to find the tawny *A. aurantiaca* in the garden take years to get rid of them.

I do not find the 'Parrot Lily', *Alstroemeria pulchella* (syn. *A. psittacina*) so eager to increase. I have had it so long that I do not remember how I got it in the first place, and it has grown on the same ledge of the rock garden for nearly thirty years. The fact that it has not much room to increase may be in his garden near Goring-on-Sea, come in shades of yellow, pink and white. I come across odd plants of *Eremurus* in most unexpected places. One day I got a message from someone in a council house in a nearby village asking if I could call and tell the owner what strange plant was flowering in her garden. Another was doing extremely well in the garden of a small brick villa, and both had the heavy clay soil of the district.

Rather unusual plants that look as though they might be difficult but are in fact quite easy are always good value. Take *Tricyrtis,* the Japanese toad lilies, for instance, which do best in shade and flower in the autumn. The one I started with was *T. hirta alba,* not very tall and with pure-white flowers. I do not think it says much for my intelligence that I chose the white form of a plant that is characterised by the strange spotting of the flowers. *T. hirta* is the most dwarf of the common species and with me is about 1 ft only. The flowers are small and a most curious shape with straight petals from the calyx and an elaborate central boss. The leaves of *T. hirta* are pale green and rather hairy and the plant has not such a Japanese look as the other taller varieties. After getting the white form I managed to acquire the type which has pale-lilac flowers and rather faint spots all over them. At first it romped away and I found odd stems coming up some distance from the main plant so I followed my usual practice of moving bits here and there in the garden where I thought they'd look nice. This was a mistake. They promptly died and the original planting has never regained its earlier abundance.

Of the three other species I grow *T. macropoda* is my favourite as I feel it has most character. It is the tallest one I know, and has flower stems usually about $2\frac{1}{2}$—3 ft high. The flowers are a deep lilac with dark and dense spottings and blotchings. This really is a Japanese looking plant with slender black stems and pointed ribbed leaves that look as though they had been threaded on to the stem at alternate intervals. The intermediate toad lilies grow to about 2 ft. *T. latifolia* has white flowers heavily flecked with purple and in *T. stolonifera* there are many mauve flowers on loose heads, all heavily patterned with purple. These are woodland plants, looking right in any woodland setting. I have not found they need any special conditions. Flowering time is late, usually about September when these strange flowers get more attention than they would at the height of the summer.

Two little plants also for a shady position which do extremely well in my soil are *Liriope muscari* and *Hacquetia* (*Dondia*) *epipactis*. I have never tried to grow them in any soil but clay so do not know how they'd react to a light soil, but judging by the way they increase in my garden, they enjoy it. Some people grow *Liriope* in full sun but my best clump is in a narrow bed facing north under a hedge, a magnificent specimen with a number of 12-in stiff spikes encrusted with small lavender flowers. A new one called 'Majestic' has 9-in spikes and deeper-coloured flowers. I like *Liriope* because it is evergreen and I know where it is growing. *Hacquetia epipactis* unfortunately is herbaceous and there is danger of forgetting where it is planted. I am always relieved when the little gold flowers appear on the surface of the soil, rather like paper flowers stuck on the ground. At first the green centres are not so noticeable but when the stalks grow the exquisite beauty of the flowers is revealed. The leaves that follow are shamrock-shaped and grow quite big—and a little coarse. The calyces remain when the petals have gone and look almost like green flowers above the leaves.

blue. But I think I like even better the delicate filigree of bare stems in winter. At one time I used to visit a street in Oxford frequently where this broom grew up from behind a wall; I do not remember it in flower but the pattern it made against the grey sky of a winter afternoon will remain with me always. In a garden I know well two of these brooms are planted to make an arch where one small garden leads to another and the effect is always memorable. I have been trying to establish it for years to grow up above my high boundary wall where I have a line of small trees. Nurseries usually supply small pot-plants and so far they have never grown more than 2 or 3 ft. Hopefully they are tied to a cane but this hasn't given them the incentive even to reach the top of it.

The variations of our wild yellow broom, *C. scoparius,* are lovely in the garden, particularly in the winter when the bright-green shoots are extra vivid. They show up wonderfully in a bed of heather and break the monotony, but they are not very happy in clay. They resent chalk, but can be grown in heavy soil if the drainage is good and if the soil is well laced with peat. I have given them up because they do not really fit into any of my pet compositions. I occasionally grow clumps of *Erica carnea,* which will grow in my soil and which has many attractive variations ranging from the brown antlered bells of 'Springwood White' to the deep crimson 'Vivellii'. It also most obligingly flowers in winter. But heathers en masse I think belong to moorlands and the only garden where I feel they fit in is one made right out on a moor. On acid soil heathers make good underplanting for azaleas and dwarf rhododendrons with touches of blue from *Gentiana sino-ornata,* but a heather garden is to me usually very dull, in spite of brooms, specimen conifers and other contrasting shapes. *Erica carnea* will tolerate lime and though it does not do well on chalk it will succeed on clay if the drainage is good and if plenty of peat is used to give the hair-like roots a good start. When planting it is most important to see there are no perennial weeds in the soil, for though no weeds from above will penetrate the thick clumps it is very difficult to get rid of such things as couch and bindweed.

There is no difficulty in growing the tall willows in clay soil, for they will do well in wet, ill-drained conditions. It is necessary only to push cuttings into the soil and they will root. Only 1 in of the 9-in shoot should show above ground. The exception to this easy form of propagation is *Salix caprea,* the sallow, which has the most beautiful yellow antlered catkins. The other cuttings can be inserted where the shrub is to grow, but the sallow is not as easy, so the cuttings have to be rooted in a frame or greenhouse.

The smaller willows are not happy in clay. They are usually found among rocks on the mountainside where the drainage is good and the soil poor and dry. I have tried several times to grow *Salix lanata,* with its silver leaves. It makes a delightful low and informal bush which looks well on a large rock garden or can be grown in a mixed border, but with me it has never looked anything but miserable. The one I would particularly like to grow is 'Wehrhahnii'. In the spring the small white catkins on its dark stems look like tufts of cotton wool, and later the plant is covered with light-green leaves. An interesting willow which does seem good on clay soil is the Japanese *Salix* 'Kurome' (correctly known as *S. gracilistyla* var. *melanostachys*). My small plants have not done much yet but the bush from which the cuttings were taken is in a garden in Shropshire on the heaviest clay soil. This willow has slender leaves of pale green and its catkins are dark crimson-brown. The bush in my friend's garden is growing sideways into a really wide clump but its height is not much more than 3—4 ft. I first met this willow in a most interesting garden in Northern Ire-

I still like pastel shades, the soft blues, lilacs and pinks which blend together so harmoniously. But a garden without some pools of deep colour would be insipid and I like to mass *Campanula glomerata,* with its clustered heads of deep blue, under the light shade of a shrub. No garden would be complete without the velvet blooms of dark-red roses but I don't want them massed in a sunny bed in the middle of the lawn. I have many trained against my old walls, and always enjoy glimpses of 'Frensham' as it flowers above *Salvia sclarea* var. *turkestanica* near the hedge at the back of the border. A little shade deepens the colour of crimson flowers. I don't grow *Dianthus* 'Thomas' in the open but against a mound of bronze-leaved fennel, which I keep cut to prevent flowering. I planted the small double Sweet William 'King Willie' under the low branches of a purple sloe but the soil was too damp and the whole patch disappeared. The rose 'Rosemary Rose', with its rich foliage and the light crimson of its flowers cries out for a little light shade.

The usual way is to grow phlox in full sun but the crimsons and flames are much better in light shade. Phlox are good plants to grow among shrubs; they look right and the flowers last a long time.

Dark-red paeonies, too, look far more intense in light shade, and so do red oriental poppies. I have a magnificent old red poppy called 'Beauty of Livermere' and by some mischance it was planted in the open instead of a little further back in the border where it would have had a little shade. This is a form of *P. bracteata* and is pure blood-red without a hint of orange. Poppies with orange-red flowers are firmly excluded from the garden but sometimes they get in with a batch of seedlings of some really delectable colour. As everyone knows, an oriental poppy once planted in the garden does its best to remain there for ever. One thinks it has been eradicated but usually a scrap of root is left embedded in the soil and one day a brazen flower will greet the unsuspecting gardener.

I admit to hard feelings about real orange flowers and try to avoid them. The double Welsh poppy, *Meconopsis cambrica flore pleno,* is orange, I suppose, but not a harsh form and I like to see it with a good green background. But the orange of *Hieracium aurantiacum* known as 'Grim the Collier' grates and is never allowed to stay. Though kniphofias are usually described as bright scarlet, light scarlet or coral, while a few are listed as orange-scarlet and others as flame, they are frankly orange. Although I prefer the yellow and primrose shades I think a few of the strong colours help the garden in early autumn but I always put them at the back and let them rise up out of a sea of green. I planted a catalpa (which I keep pollarded) to give shade to 'Prince Igor', the tallest and most spectacular kniphofia I know, and the very late 'C. M. Prichard', which has dazzling but rather stubby flowers, is toned down by the large purple leaves of *Acer platanoides* 'Goldsworth Purple'. If one has the right kind of garden the best place to grow kniphofias is, I think, in rough grass, so that each one stands alone with a green background.

Yellow can be a difficult colour too, particularly the rich egg-yolk shades, and I like to subdue it by planting with white flowers or silver foliage. I do not use any strong yellow in the garden if I can avoid it, but I do like the old-fashioned Black-eyed Susan and have a clump of one called *Rudbeckia* 'Goldsturm' between the rose 'Iceberg' and the small shrubby *Helichrysum splendidum (trilineatum)*. It blooms from July to September when the garden can be dull. The yellow of the flowers is less strident because of the prominent black cones in the centre of each flower. I used to grow *Helenium autumnale* var. *pumilum,* which is a bright-yellow dwarf and really needs a green background; I liked its lack of height but I think the dark-centred 'Mme. Canivet' (if one can find it) is a better mixer, because the petals are paler and the dark centre adds distinction. Flowers with good centres always please me. This year I was given a tall rudbeckia with pleasantly pale petals and a tall green centre. It grows to 6 ft and looks well behind the tall *Phlox paniculata,* which has heads of small lavender flowers. A striped grass such as *Miscanthus sinensis* 'Foliis Striatis' could stand at the back or the delicate *Molinia caerulea* 'Variegata' in front.

The greeny-yellow of *Solidago* 'Lemore' is welcome almost anywhere. I grow it with the tall lavender phlox, with pale-blue *Aster macrophyllus* and the little off-white *Aster corymbosus,* with its whippy black stems, and with that suggestion of green in the yellow, one could be even more daring.

I have never discovered the name of a very deep yellow—nearly orange—inula, which is short and sturdy in growth and has large heart-shaped leaves. I can't remember how it came to me but I like it because of its solid, well-fed look. I grow it in the green garden among the South American eryngiums. *E. agavifolium* has glossy light-green leaves, crinkled at the edges, and 5-ft stems covered with small spherical green flower-heads. *E. pandanifolium* which is nearby is a very late flowerer—if it flowers at all—but its narrow grey-green foliage is good throughout the year and makes a graceful 5-ft clump.

Some people grow yellow and blue together but I have never found a happy association of these two colours. I try to find a clear pastel pink, even a soft salmon pink, as a good foil for deep blue, and a really strong shade of magenta to grow with soft and pale shades of blue.

Violet and purple look well with pale pink or what the flower arrangers call 'lime green' or 'acid green', but purple foliage makes a good background for almost any pastel shade. Crimson and silver used to be a recognised theme, with silver carpets under crimson plants and crimson underplanting where silver foliage is massed. Of late years I find myself planting white and silver together, and glaucous leaves under crimson or purple. Recently I planted a purple-leaved New Zealand flax (*Phormium tenax* 'Purpureum') in a friend's garden, which has the same kind of soil as mine, and put *Othonnopsis cheirifolia* round it. I have always liked the blue-green of its leaves, but have never understood why it is called *cheirifolia,* which means that it has leaves like a wallflower. It hasn't. Discussing the plant in the garden with a doctor friend, who is a Greek scholar, I gave this name (probably wrongly pronounced) and he replied 'Yes, the leaves are shaped like a hand'. I see that 'cheirofolius' has just this meaning and I wonder if the two words have become confused and we should be calling the plant *Othonnopsis cheirofolius*. There may be, of course, a botanical reason for the present nomenclature.

A WHITE GARDEN

There has always been a liking for borders and gardens planted in one colour, particularly white. The lovely white garden at Sissinghurst Castle and the white borders at Crathes Castle and Glyndebourne have been a great inspiration to many people.

Even in the smallest garden it is possible to have corners or sections of borders planted in white. I have never done this as I think it would be too sophisticated for my cottage-type garden. The cottagers grew plenty of white flowers but they mixed them up with other colours and so I restrict my white plantings to odd places here and there, which, if they were all assembled together, would make a good white garden.

In one place a group of Pacific delphiniums dominates the scene. They are planted on the highest level of the terrace and catch the eye the moment one enters the garden. Nearby I have a big clump of *Libertia grandiflora* which has evergreen grassy foliage. Its white flowers on stiff stems are not very showy but they last a long time. Conveniently the white form of our native mallow put one of its many offspring in the group, which is held together by the frothing silver foliage of *Artemisia* 'Lambrook Silver', growing at the front of the bed and on top of the low supporting wall. In another place the rose 'Iceberg' is growing near a silver variegated philadelphus and I have planted the small white hardy *Gladiolus* 'The Bride' to come up among them. Variegated bugle, which never overdoes its job as a modest ground cover plant, is good where bulbs and corms are planted.

I find the variegated form of apple mint, *Mentha rotundifolia,* makes a soft cushion of white-and-green foliage for growing with stiff or angular plants. Another variegated plant—an evergreen this time—is *Euonymus fortunei* var. *radicans* 'Silver Queen'. It grows very slowly, increasing by underground stems, and is always neat and unblemished, with slight tinges of pink in the winter. Both these plants help a white scheme.

All members of the anaphalis family are ideal for white borders or gardens. I like best the neat *A. triplinervis* because its cushions of silver leaves are evergreen. This plant does not run like the others and has to be increased by division. I leave its flowers on the plant until late in the year as they keep their ivory colour well. The other two are taller and fill up more space. *A. margaritacea* is not so silver as the others but its pale green leaves are attractive with darker foliage, and it makes a pleasant smother round the tall and stately *Campanula latifolia macrantha,* or with a white-flowered shrub such as philadelphus or eucryphia. It needs watching as it infiltrates other plantings without conscience. I sometimes plant it at the edge of a low wall so that its 2-ft stems foam over the top. *A. yedoensis* is even taller, usually about 3 ft or more, and it flowers later. It has the typical anaphalis flowers but the leaves are willow-like, grey and lined with silver—lovely when tossed by the wind. I seldom cut them down till the spring as they are particularly good when rimed by frost. In one place where I grow this plant there is a mound of perennial candytuft, *Iberis sempervirens*. This was put in before we had the more showy varieties, but, commoner that it is, it is a fine sight when in bloom. The newer types are called 'Little Gem' and 'Snowflake'. Behind the anaphalis the sword-like leaves of *Iris ochroleuca,* with white flowers that continue for several weeks, make strong contrast. In the autumn I cut off much of the foliage of this iris as it gets yellow and untidy and, as the anaphalis holds the eye, the effect is not too bare.

All forms of *Campanula persicifolia* are useful because they flower for months on end and are graceful in growth; I grow all the white varieties I can find.

In a white garden there is scope for quite dramatic plantings. We can festoon any trees that are big enough to take them with one of several rampant white roses, such as *R. filipes* 'Kiftsgate', R. *longicuspis,* or the Garland Rose. 'Sanders White' is another double only a little less vigorous than the others. I put one on a small apple tree and it has now bridged the 12 ft or so between it and a purple-leaved sloe, and is now busily working through that. If the garden is rectangular and can take a formal centrepiece, 'Nevada' is ideal to stand alone and goes on flowering until the autumn. For a smaller garden the beautiful *Viburnum plicatum* var. *tomentosum* makes a wonderful centrepiece, especially if surrounded by stone paving or pieces of stone to make a slightly raised bed, as it is in one of the small gardens at Rowallane, the National Trust Garden in Northern Ireland. The form 'Rowallane' is rather more upright and compact than some and I use it quite successfully near a hornbeam, with alpine straw-

grew many unusual and tender plants and her correa was about 5 ft high. After six or seven years mine is little more than 1 ft but it is now producing small hanging bells of alabaster green. Other shrubs I grow are *Phlomis chrysophylla* which differs from *P. fruticosa* by the yellowish tinge of its sage-like foliage, but it has the same yellow flowers in whorls up the stems. A *Pittosporum tenuifolium* which I planted in this garden as a very small plant has survived and prospered until now. It is far too big for the place where I put it and I take every opportunity to cut pieces for the house. The small crinkled leaves on the straight black stems always remind me of expensive restaurants and I simply could not chop bits off the plant and throw them away. I know this shrub is safe out of doors in many places in the west of England but I have lost plants I have tried before in corners that were more sheltered than the open position this specimen occupies. I have never tried the green and silver 'Silver Queen' but I understand the 'Tricolor' form of pittosporum is hardier. Its white-and-silver leaves are rimmed with maroon.

I used to grow the graceful *Danae racemosa* (*Ruscus racemosus*), commonly known as the Alexandrian laurel, in this garden until my better nature prevailed and I moved it to a place in dense shade where it is much happier and increases most satisfactorily. This plant has arching stems crowded with very glossy bright green leaves. I suppose it must have flowers but I have never noticed them.

As yellow is permitted to take the chill out of a planting that might seem too restrained I have one or two small plants of the golden-leaved *Philadelphus coronarius* 'Aureus' (as well as the less usual variegated form). A small shrub I have always grown is *Cassinia fulvida,* commonly known as 'golden heather'. It is a sturdy plant with stems crowded with fine bronze foliage and small clustered flowers in warm ivory. It needs to be cut back after flowering as it has a tendency to get lanky and then loses all its charm.

There is no difficulty in finding herbaceous plants to grow in such a garden. There are, of course, the alchemillas, a genus that should be in every garden. People complain that *A. mollis* seeds too much. I admit it is a generous seeder but I would rather have that than face the possibility of losing it. I like to mass it in corners of beds so that the sprays of tiny green flowers froth over stone paths. The brown leaves are not very attractive in the winter so I cut them off, knowing that new little pleated fans of grey-green will unfold the moment the weather is kind. I also grow *A. vulgaris,* the common 'lady's mantle', which grows more often in the north and particularly in Teesdale. A friend from that part of the world had it in her garden and sent me a plant. It seeds itself about in a modest way and I collect the seedlings and grow them together. The leaves are similar in shape but more green than grey and the flowers are similar but in shorter sprays and in a slightly deeper green.

A. alpina is one of the most attractive little green flowered plants. It too is a rare British native, sometimes found in Ireland. It is fashioned on a smaller scale than the other, with small rather grey leaves divided into five to nine narrow segments. They are covered underneath by silver silky hairs which make a suggestion of a fringe round the edges. The flowers are tiny in pale grey-green and grow loosely on 3-in stems. Very often *A. conjuncta* is sold for *A. alpina.* In this plant the lower parts of the sections of the leaves are joined together. I was recently given another alchemilla called *A. erythrosora,* which is a congested little plant, with small leaves and clustered flowers on short stems. A species from Turkey has also been added to the collection but it is too soon to say much about it. Both it and *A. erythrosora* seem to be more deciduous than the others, which always leave a nobbly crown with latent buds visible in the winter.

Bupleurum angulosum is a delightful small plant with sea-green flowers. It has narrow leaves and most unusual flat flowers with pin-cushion centres. It used to grow and flower for me with great vigour but latterly it has not done well, producing few flowers and not increasing at all. I have lost it once or twice but always try to replace it because I have grown it successfully and I know I must be able to do it again.

I wish I still had *Ixia viridiflora,* which had to be seen to be believed. The flowers were a most vivid sea-green and packed against a small spike. I also seem to have lost *Aquilegia viridiflora,* but whether from carelessness or because it is a short-lived plant I do not know. The small flowers of green and brown were interesting but not very striking and the plants were definite miniatures.

I pick out the astrantias that have nothing but green on their white flowers—no suggestion of pink. *A. major* and *A. carniolica* as grown in gardens are so alike I no longer try to keep them separate. *A. carniolica rubra,* with its crimson flowers, is evidently an extreme form of *A. major* which crosses with the green forms to produce a range of flowers with pink on the petals (they are really bracts) and among the stamens. Even the large form of *A. major,* with distinct green markings on each of the bracts, no longer keeps its chaste green-and-white colouring but appears in new places with the unwanted dusting of pink. I have tried to find a secluded corner where I can put a piece of this authentic 'shaggy' form of *Astrantia major,* in the hope that it will seed itself and come true.

A drift of *Rheum alexandrae* I have planted in the green garden may or may not succeed. This is a plant comparatively new in commerce and to date I have found it extremely difficult to grow. It is not my soil that it dislikes because it does extremely well in similar limy clay in a Shropshire garden where the clumps, with their leathery leaves, increase well. It is quite unusual and most effective when in flower, with 3-ft stems hung with greenish-yellow papery bracts. Most rhubarbs do well for me and after failing to please it on raised beds, where I thought it would show up very well, I have now put it in an ordinary flat bed, in not too dry a position, and await results.

Another rather unusual plant that is much easier to grow well is the double buttercup, *Ranunculus speciosus flore pleno,* which has a most distinct green centre to its large glistening yellow flowers. I was first given this plant as *R. bulbosus flore pleno* but I soon discovered this name was incorrect. *R. bulbosus flore pleno* is a very rare plant, and mine has not bulbous roots. I grow two other double buttercups, *R. acris flore pleno,* which has clusters of small flowers on 2-ft stems and *R. repens flore pleno,* a creeping dwarf, with double flowers more in proportion to its size. Neither of these has a green centre and do not qualify for the green garden. It is difficult to know where to plant them. They look just like wild buttercups and are in constant danger of being exterminated by zealous weeders, whereas, *R. speciosus flore pleno* has quite distinct foliage.

There are two very good bulbs that I am trying to encourage as a feature in the green garden. Many people grow the giant snowdrop, *Galtonia candicans,* which is so useful among summer flowers, but the smaller pale green version is not so well-known. *G. princeps* is a most delightful plant, with single bells about the size and shape of a hyacinth, but in alabaster green. They grow better in some gardens than others mainly, I think, because they are so popular with slugs. In light sandy soil slugs are not such a menace; I have to surround my *Galtonia prin-*

ceps with slug bait as soon as the tips of the leaves begin to show.

Luckily *Ornithogalum pyrenaicum* does not seem to be such a delicacy to slugs. It has a slender spike of tight-packed small flowers of palest green, with darker markings. There is nothing bright or showy about this plant but groups of the 18-in spikes are pleasant among low-growing plants. I have various ivies I use as ground cover in this garden, including the tiny 'Plume d'Or', the tight-packed 'Minima', and various small variegated types. The bulbous plants grow through them.

Another ornithogalum, *O. nutans,* is not green but its neutral colour is not out of place among green flowers. It looks almost like a silver hyacinth and I cannot understand why more people do not make a feature of it. A dozen bulbs cost only 3s 9d and they fit in anywhere. It is one of the plants that visitors to the garden always notice. I have heard that it seeds rather too much in some gardens. I wouldn't mind that. My regret is that it has never seeded for me.

Although *Allium siculum* isn't green there is an olive-green tinge to the ivory of its hanging bells. Inside, of course, there are exciting maroon markings, but these do not show unless the plant is grown above eye level.

The green-flowered hellebores are of course well-represented because they add interest during the winter. *H. corsicus* makes quite large clumps about 2½ ft high and has flowers from December to May. I also grow here *H.* x *sternii,* the hybrid between *H. corsicus* and *H. lividus*. It looks more like *corsicus* than *lividus,* but I have seen strains where the leaves and flower bracts have signs of the pink tinge that distinguishes *H. lividus*.

In contrast with the rather pale grey-green of these two, the dark foliage of *H. foetidus* stands out well. It too opens its pale green flowers in December and carries on till May. I also grow the Italian form of *H. viridis* in this garden. It is different from our British native, *H. viridis occidentalis,* being taller, with lighter, more apple-green flowers. *H. cyclophyllus* is here too. It is one of the later types to flower, with the habit of *H. orientalis* and rounded medium-green flowers.

Odd corners of the garden are filled with *Tellima grandiflora,* one of the most beautiful of evergreen ground cover plants and pretty when the 2-ft spikes of green flowers open in May. There is also *Heuchera viridis,* which is rather a dull plant in comparison, because the leaves, although similar in shape, are dark and have no black markings, and the green flower spikes do not develop the pink shading on each hanging flower, which adds to the charm of tellima.

The only green-flowered ground cover plant I know is *Mitella breweri.* It likes to work in shade and makes a close cover of small dark leaves with short sprays of tiny pale green flowers. It is not a plant for wide spaces, but useful to cover moist ground between a few special plants. For bigger spaces in the garden I use *Luzula maxima* (*sylvatica*) 'Variegata' and the horizontal *Lonicera pileata,* which has flowers composed mainly of stamens.

Grasses add greatly to the interest of a garden like this and the tall *Miscanthus sinensis* 'Gracillimus', with narrow grey-green leaves, and the variegated form of miscanthus are beautiful. The very tall and robust *M. sacchariflorus* is a good background plant. It does not run and its wide grey-green leaves flutter and rustle like a bamboo. There is a golden form of this but it seems more difficult to cultivate and is not often available, although listed by some nurseries. I use the golden form of *Alopecurus pratensis* as ground cover, as I think it looks better when kept rather short, and this is the way I think one should treat 'gardeners' garters', *Phalaris arundinacea* 'Picta', if there is not time to keep grooming it. The stems are lax and the leaves rather thin so that it soon gets untidy and needs regular attention. A very attractive rush that makes neat tufted clumps of grey-green and carries ivory flowers on 15-in stems is *Luzula nivea*.

After the rather pale colouring of the luzula the bright green of *Cyperus vegetus* makes a welcome change. This is a member of the sedge family with typical flowers the same colour as the leaves.

Atriplex halimus came from Margate and it seems to do quite well inland as well as by the sea. Sometimes called the 'Salt Bush' because it is enjoyed by cattle in dry, bare places, it makes a low, spreading grey bush which gives shade for a carpet of a small green ivy, *Hedera helix* 'Sagittifolia'.

There is no time for many annuals but there are a few I use every year. The green flowers of *Nicotiana* 'Lime Green' go on until nearly Christmas, and here and there patches of *Zinnia* 'Envy' look well. Once you grow *Limnanthes douglasii* you have it for ever. Each year there is a great patch of its rather yellow-green seedlings, followed by white-and-gold flowers, which give it its country name of 'Custard and Cream' It is soon over and the space can be cleared. I never know how it manages to sow itself so quickly.

In shady places under large perennials or against the north wall, I have my precious green primroses, and am trying hard to find a place where the green *Viola* 'Irish Molly' or 'Dirty-faced Molly' will settle down and increase. It does well in some gardens but is difficult to keep in others.

8

PLANT ASSOCIATIONS

Much of the success of the garden depends on the way the plant material is used. No two parts of the garden are the same and they all need different treatment. Colour schemes are a matter for individual taste but colour is not everything and it is worth taking trouble to consider the different types of plant that are improved in appearance by growing together. Cherished schemes do not always work out, for some years one plant may be specially late in flowering and the contemplated companion early, so they are not out together, but gardeners get used to such disappointments and hope for success in the next season. We learn to be patient and anyhow there is great pleasure to be gained from anticipation. It is largely a matter of trial and error but it helps to read as many gardening magazines as possible and to visit other gardens. Not only does this introduce us to new plants but it reminds us of many we have known but have somehow forgotten, and enables us to see different ways of growing them. The keen gardener is never without a garden notebook because it is easy to forget all the things we see.

WINTER

The beginning of the year is, I think, the time when it is

most important to have harmonious groupings in the garden. As the year goes on there are more and more plants in leaf and flower and it is not difficult to tone down or alter schemes by cutting off flowers, or even removing plants because there are plenty more to take their place. But not in the winter. You can't play chess with your plants in mid-winter and it is not the time to sacrifice any plant, flower or leaf that gives colour or interest to the garden.

Many of the most exciting winter effects come from skeletons, the filigree of bare twigs and unusual silhouettes. I always enjoy the contorted hazel, *Corylus avellana* 'Contorta', far better in the winter than when its twisted stems are hidden by leaves. It needs to be grown in an open position so that the whole bush can be seen without interruption from other plants. I have a particularly good form of *Senecio laxifolius* with extra large leaves and at one time I set out some small plants of this round the hazel. Of course when they got too big they had to be removed, but in their early stages the effect was good. A small variegated ivy such as *Hedera helix* 'Silver Queen' or 'Harold' would look well. I find 'Glacier' produces almost white leaves in dense shade and gets greyer and more opaque as it comes into the open.

I planted hardy cyclamen under one of my twisted willows, *Salix matsudana* 'Tortuosa', and chose *C. neapolitanum* because of the luxuriance of the marbled leaves in winter. Sometimes this cyclamen is called *C. hederifolium* because many of the leaves are the shape of ivy leaves. If one takes the trouble to study them it is interesting to see how many shapes and markings there are; no two are alike and all are beautiful.

Probably most gardeners would not allow teasles (*Dipsacus fullonum*) in the garden because of the prolific way in which they seed, although most people agree that they are beautiful at all times, when the cones are soft lavender and little pools of water collect where the leaves branch from the stems, and in their skeleton form in winter, silhouetted against the sky. I think one must have some dramatic effects in winter and teasles give architectural interest if they are put in the right position. I leave a few of the many seedlings in the higher part of my garden so that they stand out above a wall. On the other side of the wall the cream and green *Euonymus* 'Silver Queen' is growing against the wall with the non-flowering *Stachys olympica* (or *lanata*) 'Silver Carpet' in front. This stachys has much silkier leaves and stands up to winter wet better than the ordinary type. It is surprising how long the teasles keep their beauty in spite of the buffeting they get in this exposed position. I think they would go on two or three years if I let them.

Another good winter silhouette is *Corokia cotoneaster,* the 'wire-netting bush'. This is a slow-growing New Zealand bush that wants an open position so that the tracery of its tortuous twiggy branchlets can be seen. It has small purple leaves and tiny yellow flowers followed by red berries in June and July and I like to see it perched on the sky line with the ground below covered with leaves that show up well in winter. Nothing is better than the bright silvered foliage of *Lamium galeobdolon* 'Variegatum' but it must not be allowed to get out of hand and any tendency to climb into the shrub must be discouraged most firmly.

This lamium gives a finishing touch to one corner of the garden which always pleases me in winter. It is a triangular bed made between two walls facing west and north, and a *Jasminum nudiflorum* with leaves variegated with gold has climbed high on the west wall. In front an *Abutilon vitifolium* keeps its foliage well and adds soft leaves of grey-green as background for a huge plant of the honey spurge, *Euphorbia*

be supplied by leaves, of all colours, shapes and textures. Some of the happy groupings of colourful leaves are contrived, but often they just happen, because if several outstanding plants grow together they cannot but look well. It was an accident when I planted *Thuja occidentalis* 'Rheingold' with *Othonnopsis cheirifolia,* and the silver-leaved *Anaphalis triplinervis* on one side of the path with a large clump of the gold variegated *Salvia officinalis* 'Icterina' on the other. Another small planting was also a lucky chance and happened because something tall was needed beside some stone steps leading to the bottom of the ditch. I chose the variegated form of *Iris foetidissima* with its bold cream stripes, because it always looks as sleek in the winter as a well-fed cat. *Arum italicum* was growing in the same bed, also a large clump of *Epimedium* x *rubrum,* which has beautiful leaves throughout the year, green flushed with crimson when they are young, and crimson-brown in winter.

Rosemary, unfortunately, is not completely hardy everywhere and even in Somerset can be killed in a very severe winter, but I always try to have cuttings in the frame and when my plants are much too big for the places in which they grow, a hard winter is really a blessing in disguise. At the moment I have a delightful pink rosemary with the tricolor sage, *Salvia officinalis* 'Tricolor', on one side and a neat bergenia with rounded leaves and pink flowers, which the local nursery christened 'Lambrook Pink', on the other. The sage has purple leaves splashed with white, pink and occasionally green and looks well with the grey-green of the rosemary.

Also in full sun, but among stones on what was once a rock garden and now has mostly small shrubs and creeping plants, *Helichrysum siculum* is a buxom white-foliaged bush in winter, and stands out against the pale green of *Hebe salicifolia* and the soft heather-grey of lavenders. A good form of *Senecio maritima* repeats the white accent in a minor key, with a 5-ft *Euphorbia characias* ssp. *wulfenii*'s blue-green crooks above.

The 'Exmouth' variety of *Magnolia grandiflora* deserves rather special plants near it, which are not too high and will not compete with its bold glossy foliage. The leaves are seldom still and show the tomentose under surfaces, which make them look as though they were lined with rust-coloured velvet. This tree is so beautiful and so pleasant to look at in winter that it really should be grown with nothing but grass as a background. I would like to have it in the open, but though it will grow without a wall to its back, it does best in a sheltered or enclosed garden.

Rhamnus alaternus 'Variegatus' is in no way in competition. It makes a wide bush of delicate silver-grey and white foliage, not conspicuous but completely harmonious. It has proved hardy with me but isn't everywhere, so growing it in the shelter of the magnolia has a dual purpose. On the other side of the magnolia I am growing an interesting little spindle from Western China. *Euonymus cornutus* var. *quinquecornutus* likes shade and, as it will take a long time to reach 3 ft, it will not crowd the magnolia. It has slender stems and narrow leaves and when it is old enough will bear large green horned fruits, which open to show a brilliant lining and hanging orange seed.

I have always enjoyed a prostrate juniper, *Juniperus chinensis* 'Expansa Variegata', which has white tips at the ends of its dark branches. I have never seen it looking better than on a sunny January day, when colour contrasts were particularly noticeable. The conifer grows against a small lead figure and is backed by several large plants of *Helleborus foetidus*, with pale green flowers above leaves even darker than the juniper. A low mound of golden sage and a very buxom clump of the sturdy *Iris foetidissima* in the background completed a pleasant winter group. In another place I have planted the juniper under a large *Cornus alba* 'Spaethii'. It makes a good background for the red stems when they are bare in winter, and shows up the bright yellow-green and gold foliage during the summer.

A rather pale and hazy effect can be achieved by planting together *Hebe cupressoides,* which looks like a dwarf cypress with foliage shaded from blue to green, *Olearia mollis,* with pale stems and small whitish leaves, and the dwarf rounded *Hebe colensoi,* which has pointed grey leaves. The olearia and hebe have white flowers, those of *H. cupressoides* are very pale lilac, and this shrub and the olearia have elusive perfumes, the first of incense and the other of musk. I am often asked by visitors what it is they can smell in this part of the garden and it took some time to track it down.

In contrast to the ethereal effect of the silver plants a golden corner gives a suggestion of sunshine. There are two small bronze hebes which are cheerful in winter, and both look rather like conifers. *H. armstrongii* will reach about 3 ft, and *H. hectori,* about 2 ft. The golden form of *Lonicera nitida,* 'Baggesen's Gold', is a taller plant if left to grow naturally and its green-gold contrasts well with the heavier bronze shade of the hebes. The ivory winter stems of a tall grass, such as *Miscanthus sinensis,* blend with the shrubs in front. In full sun the downy leaves of *Geranium macrorrhizum* are often quite orange in the winter. Sometimes they turn in the autumn but in a mild winter they colour later.

On top of the bank in the ditch garden I planted a lichen-grey *Chamaecyparis pisifera* 'Squarrosa Intermedia' with dense heath-like growth. It crouches against the bank with graceful abandon and in winter crimson-leaved *Bergenia purpurascens* gets a good background. Below *Yucca gloriosa* adds glaucous foliage, and a small rounded bush of *Sarcoccoca* is a quiet green. There are several *Helleborus orientalis* and all through the group the indefatigable *Lamium galeobdolon* 'Variegatum' threads its silvered trails.

Why is it, I wonder, that the ordinary green-leaved *Euonymus fortunei* var. *radicans* is so seldom grown or even mentioned? It has larger leaves than the variegated form of a pleasant almost olive-green and it is by no means rampant. I have had a plant for many years and it still keeps within bounds, creeping along the ground under a huge bush of *Hebe salicifolia* 'Spender's Seedling' and mingling with the bright-green frilled leaves of the Parsley Ivy *Hedera helix* 'Cristata', which trails down from the wall behind. I always mean to make more of the euonymus, because its good evergreen leaves are worthy of a more important position.

Evergreen ferns do much to make the garden less bare in winter. I am even grateful to the cheerful glossy 'hart's tongue' fern, *Phyllitis scolopendrium*, when it sows itself next to clumps of the white-flowered *Corydalis ochroleuca* which has very blue evergreen foliage, and below is a pocket where I cosset a small, almost white, variegated ivy called *Hedera helix* 'Eva'. The large fronds of *Polystichum setiferum acutilobum* with their elaborate tracery are bronzed in winter and are magnificent next to the glossy light-green leaves of *Fatshedera lizei*, which tumbles down the bank just avoiding the large lightly-veined leaves of an arum, to which I cannot give a name. Even the delicate hardy 'maiden-hair' fern, *Adiantum venustum*, which has a bronze tint in winter, is ready to partner the pale glaucous foliage of an early Caucasian snowdrop.

SPRING

When I think of spring harmonies the first pictures I see are variations of white and yellow, as there seem to be more flowers of these colours in bloom in early spring than of the more intense colours.

Osmanthus delavayi is one of my favourite spring shrubs with its neat, dark evergreen foliage that looks rather like box. In April it is covered with scented white flowers. Sometimes, alas, a late frost spoils many of the blooms, and strong wintry winds can also turn them brown. Therefore it is safest not to put it in a very exposed position. When I planted mine it was small and well-protected by the *Lonicera nitida* hedge behind it. Now the shrub has increased in all directions and the hedge is kept low to reduce labour, so every year the danger gets greater. I grow *Doronicum cordatum* just in front of the shrub and when the yellow flowers open against the snow-covered bush the effect is very good. After the doronicum is over, the clumps of pale evergreen foliage make good ground cover. This is not one of the newer types of doronicum but an old one that I had from my mother's garden when we first came to Somerset, and I find it the best for this dual-purpose planting.

When planning one's borders it is not a bad idea to work from back to front, putting the plants that are first to flower right at the back, but not obscured by an evergreen. When there is little out in the garden they will stand out well, however far back they are, and after flowering, the plants in front will screen them and one is spared a lot of dull foliage in the front of the border. In one place where doronicums grow under a *Phlomis italica* with *Pulmonaria officinalis* nearby, I have in the front *Rosa* 'Iceberg' with *Rudbeckia* 'Goldsturm', with golden, black-centred flowers as contrast.

Another happy association of white and yellow is made with shrubs and daffodils on my small upper lawn. One of the loveliest of the early shrubs is the snowy mespilus, *Amelanchier canadensis* (also known by a variety of other names), a very hardy Canadian shrub (or small tree) which covers itself with white blossoms in April, before the leaves are properly developed. The buds are beautiful and the white flowers, among the young coppery growth of the leaves, are really lovely. The plant is sometimes also called June Berry because of the maroon-purple fruits that are ripe in June but we seldom enjoy them, for the birds get there first. But they spare me the brilliant orange-scarlet of its autumn foliage. I have the white *Narcissus* 'Beersheba' planted round this shrub with 'Copeland', which has completely double flowers of lemon yellow. Of course it is a gamble, but it sometimes pays, and the shrubs and the bulbs are in bloom together.

I have the same colour scheme with *Magnolia stellata* and white and lemon narcissi but it isn't quite so dramatic because the magnolia is still small after nearly twenty years. It begins to flower in March and carries on for several weeks. This magnolia is one of the most beautiful, with many narrow white petals, which open from hairy buds. It needs care and extra peat at planting time but gives no trouble after that. It never makes a big bush so is perfect for a small garden. From one of the windows in the house the magnolia is in front of the amelanchier so that in spring there is a handsome white and gold display for many weeks.

Interest in that area is continued when *Cornus mas* 'Variegata' comes into leaf. This small lawn was a tiny orchard when we bought the house in 1937 and over the years the old apple trees have blown down. They weren't much good, except for a 'Blenheim Orange', but the blossom was lovely in the spring and I grew shade-loving plants under them. Each time I lost an apple tree, I had room for a new shrub and I think the Cornelian Cherry, *Cornus mas* 'Variegata', is one of my favourites. I have heard the small yellow flowers that open on the bare stems compared to golden powder-puffs. The buds tantalise us for weeks before they eventually open and the flowers last as long. I enjoy even more the white and grey-green foliage that furnishes the shrub till late autumn. Another pleasant surprise is the crop of big red berries in autumn, which the birds don't appear to like. An encircling carpet of yellow violets (*V. odorata sulphurea*) is all this needs.

It is not always realised that there are several other good varieties of *Cornus mas*, in addition to the type with green leaves and the variegated form. The tricolor type, *C. m.* 'Elegantissima', is lovely, with leaves a subtle blending of green, yellow and coral. My specimen is still rather small and I do not expect it ever to get very big, but the one I know best is a medium-sized informal bush against a background of green. In *Cornus mas* 'Aurea' the leaves are suffused with yellow and var. *macrocarpa* has particularly large red berries.

Another vacant space is filled with a most colourful sycamore, *Acer pseudoplatanus* 'Brilliantissimum', a small mop-headed tree with most beautiful foliage. In its early stages it is a mass of bright coral-pink and cream and as colourful as a bunch of flowers. I use the pewter-grey *Artemisia pontica* to fill the small round bed in which it grows, as anything colourful would detract from the glory of the young leaves. I used to think the tree on my lawn, *A. pseudoplatanus* 'Leopoldii' was colourful but it is almost dull beside 'Brilliantissimum', although the colours are similar.

The beautiful Japanese maples are not, as a rule, for people who garden on lime and it had never occurred to me that I could grow one until the late garden supervisor of Dartington Hall, at Totnes, gave me a small one after I had lectured there. Usually in such circumstances I give away the forbidden fruit for there is nothing more greedy and wasteful than trying to grow a plant that you know will languish and die in your soil. I don't know why I was mean enough to plant the beautiful red-leaved maple in my wall border but I did and it grew and so far seems satisfied with life. The plant had no name so I do not know which variety of *Acer palmatum* it is or whether I happened to be given an unusually tough specimen. It has the shelter it needs from the high wall behind the border and although I always use peat for planting, it has had none since. Beside it is a big clump of *Euphorbia polychroma,* a mound of green-gold in spring, and in front between the mowing stones one of the few plants of the lemon-flowered *Alyssum saxatile* 'Citrinum' which I am able to keep in my garden. Before the euphorbia has turned to green a large clump of particularly tall polyanthus comes out to complete the picture. The colour is a warm crimson and it looks like old velvet.

Leucojum aestivum is a joy in spring, when often some of us have too much yellow in the garden. It is called the 'summer snowflake' although it comes out at the same time as the daffodils; in fact I often plant it where I think big drifts of large golden daffodils need the relief of small white flowers, tipped with green. I also grow it with my favourite white daffodil 'Beersheba' against a north wall in the front garden.

This snowflake looks best grown against a wall, a lump of rock or even, as in the case of the big yellow daffodils, with its back against a tree. For years I have had a clump growing in a narrow bed at the edge of the path that leads to the house. The path is about 1 ft above the lawn and the idea was that the lush foliage of the snowflake and its arching stems of flower would hang gracefully over the path and the supporting wall. But it rarely works out as intended. The stems of this plant are rather fleshy and have always seemed to me to be too thick for the small white flowers that dangle from their tips, and it needs only a bad wind, a careless cat, dog or human to break or bend the stems and spoil the effect of the planting. It looks far better where I grow it arching from a wall with a planting of *Helleborus olympicus*. This Lenten Rose has white flowers tinged with green and it flowers early and late; in fact in some years there are a few flowers in bloom all through the year. A clump of bergenia in front and plants of *Helleborus corsicus* and *Euphorbia characias* finish the bed.

The rampant *Lamium galeobdolon* 'Variegatum' makes a good carpet under silver birches and is loose enough in growth for the daffodils to grow up through it. I planted the daffodils but several plants of E. A. Bowles' form of *Helleborus foetidus* have put themselves near the front of the planting with very good effect. Now I am trying to establish the giant snowdrop, *Galtonia candicans*, against the hedge that borders the path. I have been looking for the right place to grow this chaste and stately plant ever since I saw it looking really beautiful growing in large clumps in box-hedged enclosures with a high yew hedge behind. The hedge I have to offer is nothing but the plebeian *Lonicera nitida* but it is dark green and this is a shady corner so I think the effect should be good when it flowers in the summer.

I have always had a plant of the low-growing *Cytisus* x *kewensis* on a bed that is level with the top of a wall where there is a massive mound of the purple-leaved sage, *Salvia officinalis* 'Purpurascens', behind. A stray seedling of *Euphorbia characias wulfenii* beside the path adds love-bird green flowers and glaucous foliage to the composition when viewed from a distance. I put another Kew broom under a small *Cornus controversa* 'Variegata'. This cornus is a slow-growing shrubby tree with an upright stem and flat spreading branches. The one I managed to buy from a local nursery had its branches on one side only and matched the triangular corner of the bed where I wanted to plant it. The broom comes into bloom in May and after that a planting of Mr Bowles' golden grass, *Milium effusum aureum*, carries on the colour scheme.

I have a few slow-growing perennials nearby, the ochre-coloured form of *Lathyrus luteus* known as var. *aureus* or *L. aurantiacus*, *Coreopsis verticillata grandiflora*, and in the front I hoped to have a very attractive small plant, *Chrysogonum virginianum*, which has bright golden flowers for months on end. I had grown it before and thought its disappearance was due to neglect on my part but recently I discovered in a catalogue that it doesn't like lime. It didn't look at all like an ericaceous plant but had the hearty, easy-going look of a plant that would do well on lime. Now I shall have to think again; I could use the creeping *Waldsteinia ternata* but I'd have to watch it. It has evergreen leaves not unlike those of a buttercup and sprays of golden flowers all through the summer. The foliage colours well in autumn. Unfortunately the spot is too shady for *Urospermum dalechampii*. This would suit my purpose well because it produces its lemon-coloured flowers with their quilled petals all through the summer.

In late spring I always enjoy *Polemonium carneum* with its flesh-pink flowers. This plant has a looser and more informal manner of growth than the other polemoniums. It grows to about 10 in and looks attractive under a bush of double coral-flowered chaenomeles. I grew this bush from a cutting and have never had a name for it. There are several clumps of an off-white, almost cream, tulip growing near which blend well with the planting. The tulips (I think they were called 'Niphetos') were among the first things I bought for the garden and they have been in the same places ever since. If tulips are planted about 6 in below the surface they are usually safe from disturbance.

Another polemonium that flowers for several weeks in late spring is 'Lambrook Pink'. Actually it is a mauve-pink but when it came up in the garden I already had the taller one which appeared in the same way and which I had called 'Lambrook Mauve'. This makes a nice bushy plant about 18 in high and flowers on and off for most of the summer. 'Lambrook Pink' is a form of *Polemonium reptans* and makes a solid mass of roots which deter all weeds. It is out at the same time as *Tellima grandiflora* and the 18-in spires of small green flowers look charming behind the polemonium. The tulip that enhances this association is the soft pink 'Clara Butt' and nature usually fills in any gaps with forget-me-nots. I have been trying for years to get rid of the ordinary forget-me-not and have only *Myosotis dissitiflora,* which has bigger and more open flowers and is perennial. But the ordinary one is not so easily dislodged. Every year I am irritated by the seedlings that come up all over the garden but I still have not learned to pull up and throw away these pretty little plants which I know would look attractive if I took the trouble to replant them in groups. Those that bring the touch of blue to the polemonium are an example of the way they should be used.

SUMMER

High summer is the easiest time of all to create plant harmonies, particularly if the gardener prefers soft colours to strident ones, as I do. What could be happier, for instance, than a grouping of soft pink, lavender-blue and primrose against the dark-crimson foliage of the purple filbert, *Corylus maxima* 'Purpurea'. When possible I use plants that have a long season of blooming and I find that *Erigeron* 'Charity' produces its pale pink flowers over many weeks. *Nepeta,* of course, flowers the whole summer, and next to the erigeron I have *Nepeta* 'Six Hills', with 2-ft stems of dark-grey foliage and lavender-blue flowers. The shrubby potentillas have a long season of blooming too, and the one that completes the group is, I think, 'Primrose Dame', with large pale flowers and silver leaves.

Later in the summer I enjoy the soft lilac of *Phlox panicu-*

lata against the nut. It looks very well but the flower stems are rather long and need a mound of foliage in front. A clump of santolinas helps because the phlox is not a strong colour and it looks well with yellow flowers. My favourite companions are the greeny-yellow flowers of *Solidago* 'Lemore', which, being shorter, must grow in front of the phlox. *Achillea filipendulina* 'Gold Plate' is a stronger colour but the flat flower-heads are above the phlox so the deeper colour does not really matter. Also its grey-green foliage tones down the colour.

One of the most charming spindle trees to grow among flowers is *Euonymus planipes*. It makes an open bush about 6 ft high and has large flat leaves which turn crimson in autumn. The bright fruit dangle in bunches on slender stems and hang on longer than in some other varieties. The rather coarse species of aster, *A. macrophyllus,* grows round my shrub, and its heart-shaped leaves and sturdy pale-blue flowers show up the richness of the spindle leaves. This aster has very narrow petals which grow straight and separate round the golden discs. The green-yellow of *Solidago* 'Lemore', adds a finishing touch to a planting that pleases me, but to be successful the aster has to be kept in subjection.

Some gardeners prefer the bright gold of another *Solidago,* 'Golden Mosa', to the rather sour lemon of 'Lemore'. I agree that 'Golden Mosa' is a remarkably shapely plant and a real dazzler because leaves and stems are yellow as well as the fluffy arching flower sprays, and for me the only possible companion is a silver plant as white as possible. Because the flowers of the golden rod are small a plant with large white leaves looks best, and I would choose *Artemisia stelleriana* with leaves somewhat reminiscent of chrysanthemum foliage.

In early summer the brilliant magenta *Gladiolus byzantinus* comes into flower. Sometimes it can be a nuisance and has to be banished but I would go to great trouble to keep such a handsome plant in the garden. I imagine if one had time to cut off all the flower-heads before they had time to seed, its activities would be limited if not completely discouraged, but that would take some time when the seedlings are scattered all over the garden. Far better to collect them and grow them in groups to give splashes of colour where they can be properly displayed. Grown in this way it is far easier to stop the plants from seeding because many heads will tumble after one cut by the shears. I mass them next to an outsize *Teucrium fruticans* in the front garden, which is so happy in its sheltered corner that it has long outgrown its allotted space. Further along there is another planting of gladioli in front of *Ceanothus* x *veitchianus* and others grow with drifts of variegated honesty. I am always glad when they are in bloom because I don't mind picking them and they immediately give character to any arrangement. The petals have an iridescent look and with their white markings are very attractive. I have just discovered a source of the white form and that, I think, will look well with the gold-leaved philadelphus, *P. coronarius* 'Aureus'. This old-fashioned plant does not make a big bush and I grow one of mine with a variegated form, which does not grow very big either. They look well under the shrub rose 'Fruhlingsgold' which I have trained to make an umbrella shape and which is perfect with the philadelphus when it is covered with flowers that are primrose, shaded to white. If the white gladiolus grows as well as the magenta form it will complete this group.

Early in June *Viburnum plicatum* var. *tomentosum* 'Lanarth' is in bloom and nothing could be more beautiful than its wide flat branches which are so smothered with white lace-cap flowers that the shrub looks as though it might be covered with snow. It needs plenty of room to grow sideways so I have it under the shade of a hornbeam, which is an attractive but irregular shape. I used to grow clumps of Solomon's

Seal (*Polygonatum multiflorum*) between the hornbeam and the viburnum and they were just right arching out from the hedge behind, but when the viburnum got bigger they had to be removed as they were also increasing and coming up among the viburnum branches, spoiling the simplicity of its laden boughs.

There are several euphorbias that look well with summer flowers. *E. stricta* is a biennial with red stems and myriads of tiny love-bird green leaves. It grows to about 1 ft or so and can be planted almost anywhere for its gossamer foliage is as soft as lace. It must have flowers because the garden is full of its seedlings but I imagine they are very small as I have never seen them. One or two plants appeared near a great hummock of the golden variegated sage, *Salvia officinalis* 'Icterina', and quite transformed it. For next season I plan to add a group of white penstemon to bring in a paler note. The sage seldom flowers and when it does I remove the pale-purple spikes because I think the plant is more beautiful without them.

It is difficult to know where to grow *Galtonia princeps*, the giant snowdrop with pale green hanging bells. I have some near a white hydrangea and my best clump is by a shapely bush of *Chamaecyparis lawsoniana* 'Fletcheri' but it needs companions. Again it was a few self-sown seedlings of *Euphorbia stricta* that arranged themselves near the galtonia and added a new shade of green to the alabaster of *G. princeps*.

Euphorbia dulcis is grown, of course, for its wonderful autumn colour. This 18-in spurge has small neat leaves and like *E. stricta*, must produce flowers, although I have never noticed them. I am most grateful for all the seedlings. In late summer the leaves are beginning to turn and the effect is pretty, particularly when seen against one of the rich orange plants that flower at this time. I can get no enthusiasm for common montbretia but in *Crocosmia pottsii* the flowers are

a living host. Some of the forms of *C. heracleifolia,* with scented flowers like blue hyacinths, growing in clusters among large leaves, can look quite natural against a fairly low wall especially if a rose such as the pale-pink 'Dr van Fleet' or 'New Dawn' can be grown on the other side of the wall and trained over. The best form is 'Wyevale', and 'Davidiana' makes a good second. 'Mrs Bryden' is not as herbaceous as the others and makes long stems with pale slaty-blue flowers and leaves that are not as big as the others. It needs a warmer, more definite colour grown with it and I discovered that the light-vermilion tubular flowers of *Phygelius capensis* gave just that fillip it needed. There is an 'American Pillar' rose against part of the clematis and it looks less strident with the dull-blue flowers.

There is no discord when *Clematis rehderiana* is in bloom because it flowers at the same time as my white Japanese anemones and by luck I had planted several clumps in the same bed. The clematis now covers all the shrubs growing on the wall and some of its strong stems even fling themselves on the big clumps of *Lactuca* (*Cicerbita*) *bourgaei* in the bed in front. This blue-flowered lettuce comes out late in the year and the big heads of lavender flowers, on stems about 4 ft long, make bold clumps. They appear in odd places because the plant produces many seedlings which I find so handsome that I let them stay. The primrose of the clematis with soft lavender and pure white is a restful colour scheme. I have been censured for recommending the lactuca but I think this is because it is sometimes mistaken for *L. macrophylla,* a dangerous member of the family which runs, burrows and infiltrates to an alarming degree. This latter was given to me with some plants from a good garden and has been hounded by me ever since.

Although my favourite Japanese anemone is the single white, I find many visitors like better a deep crimson, which is semi-double and is not so tall and stately as the white. Mine may be *Anemone* 'Stuttgartia'. I have several clumps growing in the ditch garden and plant with it the late-flowering monkshood, *Aconitum napellus bicolor,* with flowers that are white dappled with blue. Another late-flowering aconitum, *A. vulparia* (*A. lycoctonum* of gardens), completes the picture. It has small primrose flowers which are borne on very long stems. In the wild it arranges itself gracefully with its lanky stems arching over other plants. I put it behind other plants so that the pale flowers rest on the shoulders of stalwarts in front. If put too near the front of the bed the stems fall over the path and get damaged or broken.

My next suggestion is for gardeners in the warmer parts of the country only. I realised I was taking a chance when I bought *Cestrum parqui* from Treseders of Truro. It was recommended as the hardiest variety and I put it against a south wall where it has thrived and doubled in size. It has the most exotic fragrance after dark and I like its pale yellow flowers against the blue of *Eucalyptus gunnii* on the wall beside it. Just behind the cestrum, the large-leaved ivy, *Hedera colchica* 'Dentata Variegata', is climbing slowly up the wall.

In the same warm position I have another border-line plant which has survived so far. *Cosmos atropurpurea* used to be known as *Bidens* and is sometimes called 'the black dahlia'. Its flowers of deepest red might be small dahlias, and sometimes look very nearly black. The centres are the same colour except for one or two golden stamens. It smells of hot chocolate. Because of its dahlia-like root, if propagated by cuttings, they must be taken early in the year. It has the blue flax, *Linum perenne,* as a companion, to show up the rich darkness of the flowers and to protect the tender plant from heavy frost. The blue of the flax is intensified in *Ceratostigma plumbaginoides,* which needs to have its roots controlled if it is to produce many flowers. It is planted between stones near the flax, and its rich blue flowers vie with brilliant autumn-tinted leaves.

THE TURN OF THE YEAR

I am sure it is not imagination that makes so many people exclaim at the depth of colour we get from flower and leaf in our limy soil. All sorts of plants turn colour in the autumn. I am always stopping to identify the odd touches of crimson I notice when I am working at that time of the year. In the summer I am sometimes impatient at the persistence of *Geranium sanguineum* with its magenta-pink blooms, but when most flowers are over for the year it is good to see the deep-red leaves of this busy little commoner. Sometimes a plant of *Geranium ibericum* produces a few crimson leaves, and if the aromatic *G. macrorrhizum* is planted in full sun, whole plants turn red.

In addition to the bergenias such as *B. purpurascens* (*B. delavayi*) and 'Ballawley' that turn a rich maroon in late autumn before they flower, some of those that remain green are sometimes brightened by an odd leaf the colour of holly berries. *B. cordifolia* produces the best colour. *B. crassifolia* is not so dramatic but with its glossy leaves the varying shades of russet and bronze are very rich. All kinds of small plants help to brighten the winter garden. *Saxifraga geum*, a near relation to 'London pride', has matt-surfaced, kidney-shaped leaves, scalloped at the edges. It always gives me a few splashes of bright crimson in the odd corners where I grow it. Being neat and evergreen it is ideal to grow beside steps, at the bottom of walls or at the edge of a shady path, where it increases steadily but does not ramp.

There are times when I regard *Duchesnea* (*Fragaria*) *indica* more as a plague than the excellent ground cover plant it is—

in the right place. Its evergreen leaves are always neat, its yellow flowers with their wide green calyces are quite pleasant, and if I were a flower arranger, I should dote on those same calyces after the petals have gone, for they have the fascination of green flowers. The fat berries that later stud the plant show up in dark places, and when in late autumn odd trails turn bronze and red, I feel quite attached to it.

The ordinary stag's-horn sumach, (*Rhus typhina*) makes brilliant autumn colour but the weather has much to do with one's enjoyment, because an early frost or torrential rain will bring those vivid leaves down in a night. I get tired of the suckering ways of this shrub and thought when I was given the less common *R. potaninii,* I should not get so much trouble. Sure enough, this small tree does not sucker as heartily as the other and has other good qualities that make it invaluable. It is shapely and has bark that is reminiscent of the snakebark maples although not so dramatic. I always think the autumn colour of the leaves is better than the common tree, and they seem to hang on a little longer. The laciniated form of *Rhus typhina* has, I think, the most intense colouring of all, not so much crimson as a luminous flame which really glows when the tree is in its full glory. The finely cut leaves have great beauty all through the year.

The more delicate shades of pink and crimson assumed by members of the euonymus family are a sheer delight. All forms of our native spindle colour well, and particularly *E. europaeus* 'Red Cascade'. It is well hung with rosy-red fruit if another form is grown nearby and I use *E. hamiltonianus* for the purpose. This has soft-pink berries and its autumn colour combines pink and orange. The Japanese *E. yedoensis* (really a form of *E. sieboldianus*) is not so often seen. It too colours well in the autumn and the rose-pink fruits hang on after the leaves have gone. I have planted three, 8 ft apart, as advised by the experts, well away from any other spindle trees in the garden. The winged spindle, *E. alatus,* so named because of its corky stems which are winged at each side, is the most brilliant in the autumn. With me it turns the brightest crimson that can be imagined. It is a slow-growing, sedate little bush, with fewer fruits than the others hanging singly and not in bunches. But though it looks so demure it becomes a ravishing beauty in the autumn and holds its leaves for a week or two. I feel sure the lime in my soil intensifies the colour. A friend, who has a beautiful garden on acid soil, gets very little colour on her bushes.

Prunus sargentii is lovely at all times, with its coppery young foliage and single pink flowers. In the autumn the leaves turn warm shades of red and crimson. 'Ukon' is another of my favourite cherries because I love its greenish flowers. It too has colourful early foliage, and pleasant autumn tints.

Cornus kousa is usually considered a shrub that needs neutral or acid soil, but it grows quite happily in my garden. I think mine may be the Chinese form, *C. kousa chinensis,* which is more amenable in my limy soil. Both types have beautiful autumn colour, but the latter is less intense.

Why don't more people grow a *Ginkgo biloba* in their gardens, I wonder? Although it eventually reaches about 60 to 80 ft it is slow-growing and suited to small gardens as well as big ones. It also grows well in London; in fact the 'maiden-hair tree' is one of the most interesting and accommodating of the conifers. The glaucous green leaves are like a larger edition of the maiden-hair fern, and when they turn to pure gold in the autumn the tree blazes with sunlight. My tree is still small, after twenty years, but there is a large one in a churchyard nearby. I always try to see it in its glory but sometimes my timing is wrong and I find only a golden carpet of leaves.

INDEX